Seat in a Wild Place

ERIK BROWN

Seat in a Wild Place

SEASONS ON A NORTHERN POND

DRAWINGS BY ELIZABETH ADAMS

WILLIAM L. BAUHAN, PUBLISHER
DUBLIN, NEW HAMPSHIRE

COPYRIGHT © 1982 BY ERIK BROWN
ALL RIGHTS RESERVED
LIBRARY OF CONGRESS CATALOGUING IN PUBLICATION DATA:
Brown, Erik, 1923–
Seat in a wild place.
Bibliography: p.
1. Bird watching–New England. 2. Wildlife watching–New England.
3. Pond ecology–New England. I. Title.
QL683.N67B76   591.9742   81-15017
ISBN 0-87233-0591   AACR2

PRINTED IN THE UNITED STATES OF AMERICA

To Emma

Acknowledgments

SINCERE APPRECIATION goes to my publisher, William L. Bauhan, whose patience and encouragement had much to do with the completion of this book. Thanks, too, to our copy editor, Anne Lunt, whose doughty blue pencil repeatedly helped to clear paths through the thicket.

Although the direct observations were made by the author, some of the lore recounted was gleaned from the following sources: The *Life of the Pond* by William H. Amos; *How Birds Fly* by John K. Terres; *Growing Wings* by Sarita Van Vleck; The Golden Field Guide, *Birds of North America* by Chandler S. Robbins, Bertel Bruun and Herbert S. Zim, and Arthur C. Bent's *Life Histories of North American Birds.*

For those who have not been introduced to Mr. Bent and his remarkable *Life Histories,* a few words may prove of interest. Bent wrote his series form 1919 to 1968 under the auspices of the Smithsonian Institution which published them as *Bulletins of the United States National Museum.* They are now available, both in paperback and hardcover, from Dover Books at reasonable prices. They should be found in every library, but are not. I found the original bulletins available for study by the public at Harvard's Museum of Comparative Zoology. The books, according to the National Geographic Society, "contain an unbelievable wealth of information, gathered from myriad sources, on everything that was known at the time of writing on the life histories . . . of every species and subspecies of North American birds." I can testify to the readability of these reports. Each bird is treated separately and at length. The chapter, for instance, on black capped chickadees, in the volume *Life Histories of North American Jays, Crows and Titmice,* fills sixteen lively pages of information, anecdotes and observations from many observers, and reads like intimate and fascinating biography.

Finally, special recognition must be given to our illustrator, Elizabeth Adams, whose illustrations have made a major contribution to this book.

ERIK BROWN

Contents

A Discovery   13

MAY

Early Pond, *May 7*   17
The Intruder, *May 8*   18
Diggers, *May 9*   20
The Bold Ones, *May 10*   22
The Gatherer, *May 11*   24
Fishermen, *May 13*   24
Handsome Scratcher, Homely Hunter, *May 15*   25
Sunbathers, *May 16*   29
Swallows, *May 18*   30
Early Evening, *May 20*   32
Coquette, *May 22*   32
The Tease, *May 27*   33
Natrix Sipedon, *May 30*   34

JUNE

The Descent, *June 2*   39
Procreation, *June 3*   40
Kingbird, *June 5*   41
A Nap, *June 6*   43
Che-beek, *June 8*   44

Silent Hunter, *June 10*	45
Big Drummer, *June 14*	46
A First Visit, *June 15*	49
Yellow Petals, *June 16*	49
Tight Fit, *June 17*	50
End of *Che-beek*, *June 19*	51
Beauty, *June 21*	52
The Odd One, *June 23*	54
He Kept His Head, *June 24*	56
Death Watch, *June 25*	57
Birthrate, *June 26*	58
Scarlet Woodsman, *June 30*	60

JULY

The Diver, *July 4*	61
Ventriloquist, *July 5*	62
Pointed Arrow, *July 6*	62
Mr. Peterson, *July 7*	65
Bluebirds, *July 8*	65
Territory, *July 12*	66
Sixty-Mile Call, *July 14*	67
A Chancy World, *July 16*	68
An Unexpected Service, *July 19*	69
Chickadee Friends, *July 22*	70
The Tonic of Wildness, *July 25*	72

AUGUST

Ballet Dancer, *August 7*	74
Prodigious Fisherman, *August 14*	75
A Touch of Spring, *August 24*	79

SEPTEMBER

The Baiting of a Hunter, *September 12*	81
Unseeing Hunter, *September 14*	82
Ducks and a Pair of Visitors, *September 21*	85

A Touch of Wine, *September 26*   88
Numbers, *September 27*   89

OCTOBER

Prince of Soarers, *October 1*   91
Conflagration, *October 3*   91
Weak Feet, *October 5*   92
Autumn, *October 10*   94
Six Eyes, *October 12*   94
Silent Feeders, *October 22*   95
End of the Season, *October 24*   96

A Discovery

TWENTY YEARS AGO I moved with my family to a small village in southern New Hampshire. Having lived for several years in a somewhat isolated spot, we were delighted with our new neighbors. I was astonished, however, to find that two of them, sturdy family men both, occasionally rose at 5:00 A.M. and disappeared into the neighboring hills on a *bird watch*! I, in turn, gave these fellows close observation for some time before deciding that this was simply the kind of aberrant but harmless behavior that one must tolerate from time to time among friends and neighbors. Bird-watching was to me an activity confined to those well-known "little old ladies in tennis shoes" who went out armed with field glasses and bird guides, peering and *psshhing* into bushes in an odd and eccentric manner.

Later, in my middle years, some of these ladies turned out to be good friends of my own age; they didn't seem old at all, but they still wore sneakers and *psshhed* into bushes. I still felt alarm, and in spite of many youthful hours spent in backcountry, I carefully avoided any activity that might be called bird-watching.

This attitude was due for a change, however, and it took place on a bright spring day. During a ramble in the forest, I came upon an old beaver pond. It was warm and quiet and I sat down in the yellow dead grass a few yards from the water's edge, with my back to a tree, and promptly fell asleep. When I awoke, some time later, the pond had changed. From a peaceful, solitary retreat in the forest, it had become a veritable crossroads of activity. Squeaks and trills and screeches reverberated across the water in a kind of wild music, and trees and shrubs were briskly alive with birds that I had never remembered seeing. With mounting excitement I began to realize that birds do not simply flit aimlessly from tree to tree, subject only to identification, but that their activity is purposeful and complex, and that if I sat awhile perhaps some of this behavior would become comprehensible. And so it happened that I became, most modestly, a birdwatcher.

The pond of the nap, revisited repeatedly in the following seasons, became the pond of this book. It lies in a narrow wooded valley protected by a conservation area to the south and an old military reservation to the north. The stream that feeds it is paralleled by a winding woods road that formerly serviced a group of early settlers, whose empty cellar holes can still be found in the hillsides nearby. The numerous stone walls that cross the area provide mute testimony to the open pastureland that once covered most of this region. Those fields, wrested with such prodigious labor from the primal woods, have now reverted to typical eastern hardwood forest, interspersed here and there with white pine and fir and hemlock. It is an uncommonly beautiful valley, the stream now tumbling through deep woods, now flowing quietly into one of a series of beaver ponds that lie like a string of jewels in the deep green of the valley bottom.

Our pond is an old one, deserted by the beavers several years ago after a February flood washed out their dam. The dam still partially holds, however, and half the original pond remains, the upper end covered with green grasses as the pond slowly reverts to marsh and bog in the natural cycle of beaver ponds. Most of the observations were made from a seat on a rock forty feet back from the west edge of the pond, with the old dam on the right to the south, and the stream inlet on the left to the north. The pond area is one hundred and fifty feet wide and three hundred feet long including the new grasses over the north end.

A scattering of dead trees and stumps, silvered and shaggy with age, stands around the edges of the pond. Of these the birds have their favorites for perching and nesting. And of these favorites, two are identified in the pages that follow. One, a silver-grey pole about fifteen feet high standing at the water's edge in front of us, I've called Near Stump. Far Stump, somewhat larger, stands directly across the pond. It has a three-inch-diameter hole two feet down from the top which is the entrance to a sizable cavity. All observations were made between 7:00 and 10:00 A.M. unless otherwise noted.

Two comments about this book should be made. First, this is not a book of instruction in bird lore, or identification, for which many excellent texts are available. It is a book of personal observations only, observations of behavior that could have been made with pleasure by any reader. It takes years to become an expert "birder" with his intimate knowledge of hundreds of local species, of their habits and habitats, their markings and their bewildering variety of calls. To become a beginning observer of behavior, however, one need only settle in a comfortable spot armed with binoculars and bird guide, and watch! My basic guidebook is The Golden Field Guide's *Birds of North America*, although Peterson's newly published *A Field Guide to Birds East of the Rockies* is excellent and perhaps more convenient for Easterners.

This brings us to the second point: To study behavior, one must be still. This is the secret. To try this pastime, find yourself a comfortable seat (I carry a small folding canoe chair) and sit. A half blind made from a piece of burlap stretched tentlike on sticks over one's legs and up to one's chest allows a modicum of squirming and the indulgence of an occasional itch. It also permits a surreptitious look at the bird book under the burlap, or the taking of notes. But such cover is not necessary. Sit motionless for ten minutes, or fifteen, or even half an hour. Take a brief nap if necessary to curb your impatience. And wait. Suddenly from the empty wood, or the deserted pond, there comes a call, a flutter of wings, or a loud splash. Your eyes pop open, your heart jumps, and the show is on. I hope the following narrative will induce you to find your own seat.

For those experienced birders who have picked up this book unaware of its possible misinterpretations of behavior, or inaccuracies of identification, I beg your forbearance and hope you will read on, perhaps gaining thereby a chuckle or two in reminiscence, and sharing the delights of discovery with the author.

MAY

Early Pond

MAY 7. It is early morning. A cold grey morning. The pond is quiet, the surface still as polished slate. One black duck beats rapidly up the water and disappears into the mist as I settle on my rock. For the moment there is no visible life around the pond. I am reminded of a pot of water on a cold stove that has just been turned up. Energy builds slowly but irresistibly at the bottom of the pot. In due course the first bubble of steam will break loose and energetically wobble its way to the surface. Soon others will follow.

And so it is with this early spring pond. The bright green shoots of new grass testify to the unseen energy. To the human eye, the young maple leaves, only an inch long now, hang motionless in the damp air. But the juices are coursing upward; the leaves are reaching for the blanketed sun. Each day the heat is turned up another notch. My eyes tell me, though my numb fingers do not, that soon this pot will boil. A sudden loud bugling announces that a bluejay is up. Various calls now come from the forest. I hope I will know them all someday. They increase in intensity as the mist rises until the chorus of chirps and warbles and trills becomes a full-scale woodland concert, brimming and trembling with energy and new life.

The Intruder

MAY 8. A pair of sandpipers is visiting today. They bob their way through the soggy grass at the upper end of the pond. I identify them as solitaries from their smooth whitish undersides, their dark legs, and their habit of frequenting freshwater ponds. There is something mechanical and robotlike in the manner of sandpipers, and I have paid them little attention until now. But this morning they are doing a beautiful thing. Every few minutes they flair their outer tail feathers, which are white and barred with brown cross bands, and facing each other, bounce into the air in a graceful, light-footed dance. Then one flutters off a few yards with the other in pursuit. This little ritual quickly over, they resume their bobbing and pecking at the grass. Then the ritual is repeated. Their wings are surprisingly long and slender, their flight as light and buoyant as a thistle on the wind.

Later. If in the excitement I have not misidentified him, we have just had an osprey! An angry osprey. He landed in the top of the old maple tree across the pond, his feet spread in a broad defiant stance between two branches. He held his great hooked wings stiffly wide, his head low and threatening, and uttered a series of harsh screams. His snowy undersides glistened in the early sun. He is our largest hawk, sometimes called our third eagle, and he plunges feet first into the water to catch fish. Today he acted every bit the part. He flew twice up and down the pond, screeching repeatedly. The reason was soon apparent. A fisherman, on this early day of fishing season, clambered over the dam at the south end of the pond, and the osprey flew off, shrieking his protest. The fisherman, clad in all the felt-hatted, fly-decked glory of the avid angler, oblivious to the commotion over his head, intently surveyed the silent pool, as concentrated on his element as I on mine. Then he flung his lure far out to the

middle, seeking that same bass that the osprey had no doubt marked from his high station. I watched the fisherman for a while, but how I would have liked to have seen from so close a range the high, wild, plummeting dive of the osprey!

Diggers

MAY 9. Much commotion when I got here. A flicker was hanging on the side of Far Stump, two feet below the hole near the top. Several tree swallows dived fiercely at him. He ducked quickly as each swallow swooped by, and finally flew away.

The swallows kept winging around the stump, taking turns landing and peering in the hole. The flicker, a large woodpecker, returned and again landed on the trunk below the hole. Again the frantic diving of the swallows. This time the flicker held its ground, ducking and parrying. Suddenly it scurried up the stump and into the hole. The battle was over and the swallows gave up. I can see the flicker moving around inside.

Ten minutes later. Can't believe it. I just rechecked the hole, and there's a swallow peacefully sitting there, its small, sleek head poked out of the opening.

An hour later, on a walk upstream, I found a more industrious flicker. As I looked at a trim hole near the top of a high stump, a head emerged and peered about sharply. The tawny color and flash of red at the back of the head identified a flicker. The lack of a black mustache confirmed a female. The head withdrew and reappeared with a vigorous twitch to the right. A small explosion of wood chips sprayed into the air. Back in went the head and promptly returned with another spray of chips. After several of these dispersements, the head disappeared for a while, presumably for further excavation at the bottom of the hole. Then a reappearance and a new burst of chips. This industrious mining of the stump was still in process when I left an hour later. No male was in sight.

But woodpeckers are not the only diggers of the treetops. On the way back through the woods to the pond I saw a chickadee on a branch above me. Just as I looked at him he seemed to emanate a little spray. I wondered idly if he was shaking off water, but having no other explanation I simply noted it and continued on my way. A hundred feet further on I suddenly thought, Spray? Spray? Do you suppose . . . ? I hurried back and found the chickadee. He *dee dee deed* at me for a moment, then flew down to a low stump, into which he disappeared as if by magic. I circled the stump from thirty feet away, and found a hole four inches from the top. The stump was a rotting remnant, four feet high and about four and a half inches in diameter. I unfolded my seat and sat down to watch. A second chickadee landed on a nearby twig. Out popped bird one with a large mouthful of whitish material, and flew up to a branch twenty feet away. As he landed, a little spray of wood bits floated off below him. Bird number two in the meantime had hopped into the hole to get her mouthful. And so they alternated, each bird spending twenty to thirty seconds in the hole while the other waited his turn, little puffs of rotted wood chips floating to the forest floor after each trip. After ten to fifteen minutes of this they both flew off into the woods for about five minutes, as if on a work break.

On one of these breaks I hurriedly measured the hole. It was just over an inch in diameter at the opening, two inches across inside, and eight inches deep. Now I sat down ten feet from the hole. When they returned they looked me over briefly, then the larger and bolder male hopped in and went vigorously to work. I could plainly hear his tapping as he wrestled bits of wood from the sides of his snug new home.

I don't know why chickadees disperse their chips. If it is to keep secret the location of their hole, it would seem out of character with their usual fearlessness and carelessness of safety. Also I cannot say if this hole was dug from scratch, or was only being enlarged. I suspect the former, as the hole seemed too low to have been started by a woodpecker.

The Bold Ones

MAY 10. A hoarse croak greets me this morning from across the pond. This must be a bullfrog with the belly of a whale, I think. The sound is soft, tentative, and impossibly deep. I scour the opposite shore with the glasses. Suddenly the croaks rise in sharp crescendo to something between a bark and shriek. Now a chorus of sharply resonant, high-pitched barks racks the area. They echo deafeningly, as if the pond had suddenly become the inside of a big drum. Now the owners come straight across the pond toward me. They are a magnificent pair of Canada geese. Their black heads with the sharp white underbibs are held high. One of them snakes his head back on his long black neck, then stretches it forward in a strange rocking motion. They have probably both heard and spotted me. They steam boldly to within twenty feet of my shore with din unabated. Harboring the residue of an old chest cold, I have badly needed to cough for the past ten minutes. I line up behind a tree and cough loudly with impunity. The sound is lost like a whisper in the cacaphony of these geese. They steam up and down in front of my rock, peering sharply into the trees, keeping up their irate clamor for almost fifteen minutes. These are big sounds, sounds that rank in the wilderness with the cry of loons, the deep woof of the moose, or the quavering calls of coyotes—sounds that rip into the heart and pluck mightily at ancient memories buried there.

Three black duck follow the geese out into the pond, circling nervously. After five minutes the tension proves unbearable, and the ducks explode into the air, beating rapidly off to the north. But no quick and timid departures for the geese. How aggressive and fearless they are. One often hears them passing in the dawn sky, chattering carelessly and aimiably as they head for the great barrens a thousand miles to the north. They tolerate me now, and allow me to climb onto my rock. Only a glint of sun striking from their black eyes makes the eyes visible in the blackness of the heads. The great grey bodies seem huge, and float high in the water. They are still keeping sentry, with only an occasional bark, when I leave an hour later.

Canada geese often set their nests on top of old beaver houses, lining the egg hollows with soft down. But be careful if you approach such a site. Audubon reports that one big gander attacked him so fiercely that he thought for a moment that its beating wings had broken his arm. I would have feared more a strike of that heavy beak to the face.

Audacious as they are, these geese, both young and old, can readily dive to safety, according to A. C. Bent, and swim thirty to forty feet under water. Inasmuch as they molt their primary wing and tail feathers during the summer breeding season, and cannot fly during this period, diving would seem to be a good defensive mechanism—particularly helpful in protecting the vunerable young. An equally unexpected and interesting habit is their custom of "playing possum"—lying flat on the ground with head stretched low before them when they wish to hide. This apparently is often practiced by the female on her nest, but one observer tells of seeing two geese lying in this position on a sandbar on a river, fully in the open. They looked so much like old pieces of driftwood that he would never have noticed them if he hadn't spotted them landing there from far up the river.

The pair out before me now show no such caution. Last year a pair of geese, perhaps these same two, nested in a nearby pond and raised five goslings. I hope these two retreat there—or somewhere—soon. Much as I have enjoyed their visit, they are too imperious to enjoy as close neighbors.

The Gatherer

MAY 11. The tree swallows have apparently won the battle of Far Stump. A little female is building in the hole. She makes repeated trips across the pond to a small area just below me, picking up dead pieces of grass, wresting them vigorously from the matted tangle at the edge of the water. She returns each time to the same area, not more than three or four yards in extent, although there is the whole marshy end of the pond from which to pick. Now her mate drops out of the bright sky and lands lightly on top of a nearby stump. He watches her as she makes at least a half dozen trips, chittering lightly all the while. How sleek he is. Sitting still he seems in motion, his folded wings round and smoothly tapered back to long points along the streamlined tail. His dark blue back contrasts sharply with snowy undersides. He is handsome, but he does not help with the nest building. Now he wings swiftly over to the hole, peers in briefly, then swoops off across the pond.

It is a cold, brilliant day and my fingers are stiff as I write. But the pond is cheerful. The nuthatch utters repeatedly his plaintive two-note cry; flickers call to each other in their long-drawn-out manner. Two redwings burst across the water in excited pursuit.

Fishermen

MAY 13. Blackflies are out. Two fishermen walk down the trail behind me. The low rumble of their voices precedes them and alerts me to be motionless. Fishermen and I are always affable when we meet, and each surveys the other with considerable curiosity. I, however, with my field glasses and folding seat and camouflaged hat, am the rarer species, and am consequently subject to the closer scrutiny.

The fishermen pass, but the pond is quiet today. The weather is cool and blustery, and this seems to dampen activity. The male swallow perches for a while on a twig protruding from Far Stump, but there is no sign of his lady who was so busy the other day refurbishing the nest. Three pairs of redwings periodically flutter up from the dead grass at the marshy end of the pond, but they are still flocking and show no inclination to start nesting activities.

Handsome Scratcher/Homely Hunter

MAY 15. On the way to the pond a rufous-sided towhee is singing off to one side of the trail. I never see him in the open space over the pond. He likes the woods and often can be spotted scratching through the leaves on the forest floor. This morning he sits on a branch of a young beech tree calling vigorously, head thrown back at each call and vibrating rapidly down to the tip of his long tail. He looks like a small black knight of the forest, uttering challenges to all comers. He is striking, with a coal black helmet that covers to his chest, punctuated, when the sun catches just right, by a glint from his bright red eye. He is white underneath with brick-colored patches under each wing. The long, dark tail and wings are trimmed in white.

As he sings I seem to be able to identify at least four calls in the time of half an hour. They go something like this: *Drink your teeeeea*. Then later a few brief *rrrrps*, then quickly, *brrrt chi chi chi chi chi*. After a bit the call changes to *Chip-re chi chi chi chi chi*, and this continues for perhaps ten minutes. Bird calls are difficult to describe. When we humans attempt this, we add consonants where few exist. Most bird calls are vowel sounds proceeding deep from the throat. In English we identify several basic vowel sounds, but birds seem to be capable of uttering hundreds of such sounds. We may not know the number and complexity of the messages conveyed, but for sheer beauty and variety of utterance, birds are unmatchable.

Now there is an answering call to the towhee. Does he in fact identify each of the profusion of calls in the woods? Does he say to himself in some birdlike fashion, "Now there is that pesky nuthatch again," or, "I wonder why those jays are so upset?" Or does he select and identify from the woodland chorus only those answering calls of his own kind? Are all the other calls only unidentifiable parts of the never ending melodic patter of the forest?

A green heron has just landed at the upper end of the pond. He is my first, and I am frozen with excitement! His back looks slate gray to me, not a bit green. He pokes his head up to look around. His long neck is a dark reddish color and strangely thick for the size of his body. He has a black crest on his head. Now he strikes! He has a large wiggly mouthful. It is a young frog, caught across its middle, hind legs spread and twitching spasmodically. The heron hunches over repeatedly, almost disappearing in the grass, apparently trying to line the frog up head first and aim it down his throat. Now several vigorous gulps and the frog is gone. After five minutes of near immobility the heron makes a short two-step dash and another strike. Another frog! He's deadly. More hunchings and a quick gobble and down goes the second frog. How can he conceivably eat so much in such a short time?

He flies up in a slow circling of the pond, and, to my delight, lands on a dead branch not thirty feet away. He folds his thick neck into an **S** and tucks it into his body so completely that he

assumes the shape of a crow as he settles on the branch. Where on earth did the neck go? He is strangely unwary. I think he is slightly stupefied from his heavy lunch. Now a big yawn. Then a long white dropping streams down behind him. The heavy lethal beak is longer than his head. The wing feathers are faintly outlined in yellow, and two whitish stripes run down his purplish chest. His yellow eye has a black dot in the center. His crest is a handsome blue-black, but his tail looks as if someone had just snipped it off with a pair of scissors. Long skinny toes sprout from a pair of naked orange legs and wrap themselves securely around the branch. The Great Creator had an off day when he put together this heron. He is a bundle of mismatched parts.

He yawns again, and I yawn too. My joints ache and I make a motion to flush him off his branch. He promptly hops off, makes a few flaps, and sails off in a long glide. He is a beautiful flyer. I expect him to leave the pond, but to my surprise he glides up to the top of Far Stump. A great blue would have croaked loudly and departed for the day. Our green heron perches unconcernedly on the stump, apparently unafraid of me. But the tree swallows nesting in the hole just beneath him will have no part of this. Male and female rise to the attack and dive at him repeatedly. He ducks and parries and flaps his wings to keep his balance, but refuses to move. The swallows are frantic. They swoop down at the back of his head like small thunderbolts. He stubbornly holds his own for perhaps four or five minutes before finally giving up, flapping his way off down the shore to continue his fishing, looking for his third frog steak dinner in the forty minutes we have shared the pond.

Sunbathers

MAY 16. I have several times heard a bird sound in the forest that could only be described as a loud, clear *urrp*. As I have tried to track down its source, it has receded eerily before me. Today I located its owner, poking at a nest of tent caterpillars high in a wild cherry tree. It was a Baltimore oriole, resplendent with his black head and bib and bright orange underparts. He pecked at the small, fuzzy caterpillars just emerging from their nest, finally dropping one which then hung helplessly from the branch by a spider-thin thread, but protected, I am told, by its thick, hairy coat. The oriole soon gave up on this bristly banquet and flew off. It is too bad. Most of the leaves of this tree may be consumed by these little caterpillars as they double and triple their weight in the process. Unlike the fall webworm, which builds its unsightly nests and does its devouring after the tree has completed its summer growth, the tent caterpillar does its damage in the spring, and is much more destructive to the trees. The oriole, I hope, will soon favor us with his clear flute-like warblings. He is a superb serenader.

Four painted turtles were lined up in a row on a small log this morning when I arrived. This perching on a rock or log in the sun is hard to understand from any practical point of view. They sit always with small black heads held high, looking very alert. Do they simply enjoy the heat of the sun of their backs? Surely they are not feeding. Their vigilance is surprising. If one walks within thirty or forty feet of them they promptly dive. Their heavy armor makes them so nearly impregnable that it is hard to believe that they have any natural enemies. I have counted as many as fifteen of them around the pond at one time.

I have a theory on the sunning habits of turtles that I have just had partially confirmed by William H. Amos in his *The Life of the Pond*. Turtles are cold blooded animals spending much of their time in early spring in forty to fifty-degree water. I suspect that after a night of hunting and feeding they must climb to a perch in the sun to digest their food. The sun would raise their internal temperatures as much as fifty degrees, greatly speeding up their metabolic processes, perhaps vitally. If so, it is a delightful necessity. I have for years tried to persuade my wife that a good nap after lunch serves some similar purpose. But she disagrees, and says that I cannot now use my distant relationship with the ancient order of cold blooded turtles as a justification for such a slothful habit.

The kingbirds are back! Wonderful. This pond is simply not its natural self when unsupervised by the constant attention of the kingbirds.

Swallows

MAY 18. The female tree swallow sits in the mouth of her hole in Far Stump, peering out over the water. The male sits on a twig just below her. The ceaseless flight of the swallows gives such an impression of near perpetual motion that I am always somewhat surprised to find them resting. The other flycatching birds of the pond, waxwings, kingbirds, phoebes, and others, make only brief fluttering sorties into the air, returning promptly to their perches. Swallows must use up enormously more energy in their marvelous constant soaring.

The three swallow-like birds in our area have surprisingly different flight styles. All three can be seen in the air at almost any hour from dawn to dusk over our farm. At treetop levels fly the chimney swifts. Small, dusky, cigar-shaped birds, they fly at high speeds. John K. Terres reports in his *How Birds Fly* that British ornithologists have timed some swifts at eighty miles an hour while feeding, and, although it is difficult to believe, have clocked the spine-tailed swift of India at an incredible 200 miles an hour on a measured straightaway. The flight of our chimney swifts is straightforward and determined, with many fewer abrupt changes of direction so typical of swallows. Their wingbeat is so rapid it is virtually a flutter. Occasionally they whiz down around the house, two or three abreast in perfect formation like little fighter planes. They are also accomplished gliders. I have watched them hundreds of feet up, circling in great arcs with only an infrequent vibration of their narrow wings. Perhaps they rest this way. It is said that they do not perch during the day, flying continuously, and returning to their chimneys for rest only at night.

Tree swallows are those that we chiefly see here at the pond. They are quick, agile flyers with a rapid wingbeat, often breaking into a short glide, like little blue and white sailplanes, as they come in to their nesting holes.

But if I could be a bird for a day, and lie on the wind swift and free, it's a barn swallow I would be. Speed, agility, and a fluid grace combine in these elegant black and tan birds to produce miracles of flight unmatched in any other, I think. They have a curious way of laying their slender wings back along their sides after two or three powerful beats, as if shooting their bodies through the air. Watch them now as they pump their way up through the orchard, hunched shoulders parting the air as they drive for elevation. Now back down they come, streaking low through the trees at such speeds as seem sure to doom them to disaster, whipping this way and that, seeking the erratic insect. But they are incomparably agile, and never collide with tree or each other. Watch one chase another for a piece of white feather, hurtling in and out of the barn doors with lightninglike changes of course, their long forked tails stretched tautly wide to gain maximum leverage. The air seems to be of an impossibly low density to support such swift and endless acrobatics.

It is estimated by one researcher as reported by Bent that these indefatigable flyers cover as much as 600 miles a day when feeding their young. In the fall they will travel thousands of miles south, some of them as far as Argentina, for their wintering.

Early Evening

MAY 20. At the north end of the pond the stream channel turns abruptly west at the old beaver house, leaving a gap in the forest. The rays of the setting sun pour down through this opening and lie across the pond like golden chopsticks in a bowl of jade. The young spikes of green grass have pushed up six inches above last year's matted tangle and give the effect of a sun-shimmered emerald haze across the marshy upper end of the pond. In the water areas hard green lily buds poke up stiffly, small wooden soldiers standing rigidly attentive to the vital sun. Peepers provide musical background with long trills following and blending one into another, like singers in an elfin round. And as in all fairylands, there are demons. The mosquitos are out in force plaguing bird and watcher, providing both torment and banquet in their droning sing-song way.

The male tree swallow lands on a small branch on Far Stump, caught brilliantly in one of the golden shafts. The fieldguide incredibly, says he is green. But he is blue, blue, blue!

Coquette

MAY 22. A short, sharp splash. Then another. A small female barn swallow strikes the surface of the pond in a burst of spray. Is she fishing? Perhaps feeding on one of the small aquatic creatures that scurry just above or below the surface of the pond? But there are several other swallows swooping their airy way about the pond and none are striking the water. She is alone in this odd behavior. Suddenly she flies up to a perch on Near Stump and proceeds to vigorously preen herself. First one wing is stretched to be plucked and groomed, and then the other. Now a rapid combing of the breast feathers, down across her underside and back to the tail. This takes five minutes of constant fussing. Then a quick short fluttery flight into the air. Is she drying off the last drops of water?

Now back to the top of the stump. The grooming is over. She spreads her wings and chitters excitedly. The signs are unmistakable; she is inviting attention. Freshly bathed and primped and tidied, she is ready for a visit from one of the handsome males. Is it a particular one that she calls to? Or is this her first attempt to attract a male? She calls in vain today. The feeding is good and the males do not interrupt even for a moment their endless soarings and swoopings over the pond.

The Tease

MAY 27. The edge of the pond was in a state of excitement when I got there. There was a muted flurry of wings overhead, then flashes of color. After a few moments of silence there was another flurry. Suddenly this activity broke into the open with a Baltimore oriole in violent pursuit of a bluejay. The jay led a merry chase back and forth through the treetops; then both landed in branches not more than four feet apart. Each eyed the other impassively for a moment or two, then the oriole dashed at the jay and off they went on another brief, dizzying race. At one point the oriole sat longer than usual in a nearby tree, presumably catching his breath, when the jay suddenly swooped down over his head exactly as if to provoke him. Off they went again on a rollicking tumble of bright blue and orange, with the oriole in hot pursuit. Is the oriole defending a territory? Against another species? Or is it possible that they are only two bully boys provoking each other for the sheer joy of doing battle?

The chase goes on for half an hour, circling about this patch of woods, the participants never at any time flying out of range of my sight. The only other observer seems to be a little least flycatcher, who actually follows them from tree to tree uttering his *che-beek, che-beek*. Many other inhabitants go by, however, without seeming to pay any attention. A great blue heron flaps slowly overhead. A towhee passes not ten feet from my seat, scratching with his usual zeal in the leaves. A bright little yellowthroat flits through briefly, looking like a small bandit with his coal black mask. A flicker lands on a dead branch by the pond, making an unusual croaking sound. There is simply too much to look at! The jay finally flies off, ceasing his apparent teasing of the oriole and ending the contest.

Natrix Sipedon

MAY 30. Today, as I watched a sandpiper bob his way down the shore immediately in front of my rock, my eye caught what seemed to be an old weathered piece of cable lying in the water. As I stared at it, trying to discern what it could be, it suddenly, without a ripple, *slid forward*. The hair rose sharply and eerily on the back of my neck, and I hung glued to the glasses. There was no question; it was a great snake, the largest I had ever seen in the wild.

I have several times recently watched a family of blacksnakes hunt in this exact vicinity. In fact I was treated the other day to the sight of an elegant young male following a much larger

female out of the water, up into the dry yellow grass at the edge of the pond, where after some preliminary circling, they mated, joined at the lower ends of their sinuous bodies, tail tips vibrating tensely, presumably in ecstacy, for the duration of this unselfconscious coupling in the sun.

But blacksnakes are sleek, black, and lively in the water. The large snake before me now was none of these. It was thick and heavy of body, greyish brown in color, faintly marked from this distance, and with what I could see now was a flat triangular head. My neck prickled again. While working in a Texas swamp I had once seen a large water moccasin. This snake seemed identical to the one of my memory, perhaps even larger. Could it possibly be a moccasin so far north?

I watched the snake. I watched him for fifteen minutes. He lay as motionless and uninteresting as a dead log. I became restless. I can watch this snake all day long, I thought, and he'll outlast me. Have you ever watched snakes in the zoo?

I lifted the glasses and looked about the pond to see what activity had developed in my absence. I watched a chickadee building its nest in a low stump near my rock. I looked the length of the pond. I looked into the sky. I looked back at the snake. The snake was *gone*.

I cursed my impatience, and frantically ran the glasses up and down the shore. What happened next made my eyes bulge. Where the snake had lain, there was a sudden slow roil of water, and out of this, cutting the surface like a knife, emerged a wing—a sharp, wet, tautly spread, translucent wing. Impossible, I thought; has this snake caught a bird, perhaps the sandpiper, right before my eyes, and I have missed it? I watched. Now a large loop of the snake's body heaved from the water in a slow thrashing boil. Now the head. It was not a bird the snake held, but a fish, a catfish, perhaps eight inches long, with one pectoral fin spread unnaturally wide, like a wing in flight. The scene, magnified in the glasses, seemed gargantuan and prehistoric.

The snake had the wriggling fish gripped firmly behind the head, and now undulated his way slowly toward the shore. I am still more excited observer than naturalist, and the following events bear this out. I slid off my rock and began a slow stalk over the forty feet of ground that separated me from the snake's landing place. When I got to him he was moving slowly up through the old dead grass. The naturalist deserted me completely. I grabbed a long stick, of which there are many lying about the shore, slipped it under the snake, and carefully lifted him into the air, fish and all. He made no effort to escape, and he was *flat*. He was flat, I tell you, and he hung down stiffly on either side of the stick like a great thick ribbon. It was extraordinary. Is it possible that watersnakes flatten themselves for easier maneuvering in their fluid environment? I shall have to find out. The snake slipped off the stick and fell partially back into the water, still gripping the fish. He showed no fear at all and only stared at me (balefully, I think) with one black eye peering over the back of his fish. His head was like a huge flat clamp wrapped inexorably around the fish's neck. The fish now made no motion at all. I lifted the snake again, and this time he dropped the fish. I gingerly set the snake down near a stump and glanced quickly at the shore to see the dazed fish wagging slowly off to deep water.

Now the snake's attitude changed abruptly. He coiled rapidly and struck repeatedly and viciously at my stick, his mouth held so wide that I flinched at every strike. I did not plague him, however, and soon sat down on a log nearby. He was no longer flat. His body had reassumed the thick, heavy look of a rattler. We stared at each other for what seemed an age, the heavy flat head and flickering tongue inducing in me the twinges of horror that the serpent has induced in the heart of man since the time of

Adam. After what was probably only a few moments, he uncoiled the great muscular body and without haste or apparent fear slid slowly off down to the pond again.

A book on reptiles identifies our snake as the common watersnake, Natrix sipedon, sometimes known as "moccasin," growing to an occasional length of four feet, and describes its temperament as "truly formidable when angered or cornered." It is not poisonous, lacking the fangs and the "pit" in front of the eyes of its look-alike, the southern water moccasin, a pit viper. The water moccasin is not found north, it is said (and I hope), of Virginia.

JUNE

The Descent

JUNE 2. OCCASIONALLY a sighting occurs that burns itself into the memory as if etched there with the arc of an electric stylus. Some combination of sun and smell and bird and sound that focuses all the beauty and music of the swamp into one brief, unforgettable scene. Like the day a redwing landed above me in a small dead tree of a swamp, silhouetted blackly against a blinding sun, red patch gleaming, feet spread wide on a silver branch, head flung back to hurl his harsh scream defiantly into the impassive sky, as wild and timeless as any denizen of an ancient Permian swamp.

Today's scene was nearly soundless, but profoundly equal in its breath-catching beauty and sense of antiquity. Some motion in the sky caught my eye and I lifted my glasses. Over the tops of the high trees on the far side of the pond, gliding directly at me on absolutely motionless wings, came a great blue heron. He was descending at an angle, of course, but through the glasses, and from my position directly in front of him, no forward motion could be detected. He seemed to be hung against the high verdant wall of the trees, dropping almost imperceptibly and vertically, like some huge lizard bird, some great ungainly Ichabod from the world of dinosaurs and primeval swamps. His long, knobby legs hung below him and his wings were so thin and angular from this front view that they could have been made of leather. He seemed to be looking directly at me as he came and I hardly dared to breathe. Time froze and the moment seemed forever.

At the last moment, unaware of me, he turned sideways in a smooth arc and glided with incredible grace into a soft two-step landing. Now, in profile, with wings folded smoothly against the suddenly massive blue-grey body, he seemed not the same bird of the descent. Gone was the awkward angularity of the front view. The long neck settled into an elegant **S** and the front plumage flowed down his chest like delicate embroidery. He strode slowly with long deliberate steps up through the greening grass at the edge of the pond, his head snaking back and forth with each stride. He poked among some dry sticks for a few moments, then at a sudden inadvertent movement of mine from across the pond he froze. For five minutes he made no motion. We stared at each other like two creatures fixed forever in an old print. Then he sprang into the air and flapped his way down the pond, croaking loudly and hoarsely. He disappeared over the dam, swung up behind me, and flew far upstream, croaking all the while. I would have to wait for subsequent visits to see the quiet stalk and lethal thrust of that mighty beak. And I learned that in the presence of the great blue heron one must be as still and as impervious to time as the rock on which one sits.

Procreation

JUNE 3. The robins were busy today about the business of procreation. A large, aggressive male, which had earlier defended his side of the swamp by vigorously driving off two would-be competitors, suddenly looked up from his feeding and saw, across the water, a most enticing sight. A sleek, plump female was perched on a low stick on the far shore fluttering and quivering violently, her wings held wide and low and her tail high. The male promptly flew across the pond and landed in a shrub nearby. He watched her antics for a few moments, then dropped down and mounted her. A brief struggle ensued in which both birds fluttered into the air. The female perched back on her log and started again her provocative shimmy. But the male could stand no more, for the moment at least, and fled off down the pond.

Fifteen minutes later a pair of swallows demonstrated a different stamina. A little female swooped gracefully up to the top of Near Stump directly in front of me and crouched motionless. The male swung down at her from high out of the sky, coming to a brief fluttering halt in the air a foot behind her. He dived at her as she lifted her tail high for a short decisive coupling. Then without landing he backed off, still in the air, hovering rapidly behind her, and came at her again in vigorous assault. Again he backed off, hovering, again he plunged, over and over until ten times he had consummated the act, ten times assuring the continuance of his species in the coming generation. Only then, blithe lover, did he sail off across the pond. The little female smoothed her feathers and perched patiently for ten minutes before finally flying away.

Kingbird

JUNE 5. One day, in some excitement, I pointed out to a little girl a kingbird sitting on our trellis. I had never seen one so close to the house before. "Well," she said, "I can't tell him from a swallow," her expectation of an exotic sighting worthy of the name of "kingbird" somewhat dashed. What could I tell her? They are both flycatchers who feed on the wing. Both kingbird and tree swallow are dark on top and bright white below. They both enjoy living in an old beaver swamp, and both are scrappy fighters.

But the comparisons soon end. "Look," I said, as he fluttered into the air after an errant fly, "see the broad band of white across the tip of his tail as he flares it wide in flight; that is his badge of identification."

There are other differences. The swallow is a group bird, and no one bird ever achieves individual distinction, at least to us. The kingbird, by comparison, is perhaps the most distinctive individualist in the swamp. As he sits erect on a dead twig in the center of the pond, dark coated and white bibbed, his small black square head turning this way and that, he seems the epitome of a proud, though diminutive, manager of a large and beautiful restaurant. In contrast to the constant soaring of the swallows, he makes only brief sorties into the air after insects, occasionally making a short trip down the length of the pond as if on inspection. Every so often he makes a quick, aggressive plunge into the thick marsh grass and flushes out a redwing. There follows a brief, noisy skirmish, with the angry blackbird soon driving him off. This fazes him not a whit, and an hour later he will repeat the performance.

Unlike the tree swallows, which always nest in holes, the kingbirds have boldly built their open nest in some dry, gnarly brush out in the middle of the swamp next to the open water. One of them can be seen now, head and white-tipped tail poked pertly alert from either end of the nest. She will not be disturbed, I think. Kingbirds are strongly territorial, and have been seen driving off crows, and even hawks and eagles.

If you have ever climbed a tree to examine a kingbird nest, you know how fierce these attacks can be. The birds hover directly above you and plunge to the assault within inches of your head. I have often watched barn swallows dive at cats, and small birds at crows, and crows at hawks. In each case the larger creature winced at the rush of wings, but not so sharply as to indicate an actual strike. I believe this is not so with kingbirds. They have a small, needle sharp hook at the end of a strong beak that I suspect has a purpose other than catching insects. One day I watched two of them attack a large hawk high in the sky. At each dive the hawk flinched so violently, half rolling over with such a flailing of wings, that I was sure he had been painfully jabbed on the back of his head with those sharp hooks.

Our little girl hunched her shoulders in sympathy with the hawk and agreed that the kingbird was not given his name lightly.

A Nap

JUNE 6. What a sumptuous time of year is early June. The stream still runs freshly into this half pond, half swamp. The marsh grass pushes up thick and green from its soggy soil. The water lily buds are yellow now, and lie scattered over the pond like buttercups. The air is still, the sun hot, and the inevitable wasp drones ceaselessly in the nearby shrubbery. A slight haze over the pond portends the onset of a hot day, a prediction made earlier by the weatherman. A song sparrow stops close by to sing his brief aria. Three *prrps*, a short trill and a little lift at the end. Always the same song. The dark spot on his streaked bib gives him away even if his song did not. A Baltimore oriole lands on a branch in the shadow of a nearby pine. This must be the same bird that tussled so vigorously with the jay a few days ago. Maybe he is nesting in this area after all. He preens his chest, and suddenly, in an extraordinary bit of behavior, fluffs himself out until he is nothing but a brilliant orange ball of feathers, hung like a small bright ornament against the dark pine.

On the old deserted beaver house a painted turtle has spent the last hour clambering up over the sticks until he is now a foot above the water. Why does he climb so high? What does he see from what (for him) must be such a daring and precarious perch?

By midmorning the mosquitos have left, bird activity has nearly ceased, and I take a peaceful nap on my rock.

Che-beek

JUNE 8. What a prodigious amount of energy small birds put into their calls. A little least flycatcher is probably the most persistent caller about the pond. At each *che-beek* he throws his head up sharply, his body jerking convulsively right down to the tip of his tail. It is a loud call for so small a bird and it demands a total body effort. He does this hour after hour, day after day, ranging up and down the near side of the pond. He seems to be alone, and I can surmise that he is calling for a mate. If so, it has been a

long and fruitless search so far. I shall report if a lady friend shows upon the scene.

Silent Hunter

JUNE 10. A flash of blue in the grass. A bluejay is coming down the shore, skipping his way in a series of two-footed hops from stone to log to stump. He is as agile as he is handsome. But he is not on display now. He is a hunter, quick, silent, and alert. A short jump into the grass and out again. He has a large-winged damselfly. He holds it under his foot on a log and pecks at it vigorously, gulping each morsel. Another quick foray and he has caught another. Soon, sharp-eyed and vigilant, he continues his light-footed journey down the shore.

I think I just saw a great crested flycatcher. Much larger than a least or a phoebe. Grey bib and bright yellow underbody.

Brownish back and long rusty tail. His call is a single loud *wheerp*. He landed on Near Stump, peered in the hole, and was promptly driven off by a swallow. I did not get a good look at his crest.

Big Drummer

JUNE 14. As I approached my rock this morning I heard a flutter of large wings overhead. Sudden large sounds in the forest always have particular drama, and set the heart to pounding. I swung the glasses up ever so slowly and focused, to my astonishment, smack on a pileated woodpecker. He immediately flew off, and I sputtered in disappointment. But a few minutes later I caught him on a high dead branch a hundred feet up the pond. He is as big as a crow, with a large dullish black body, as if clothed in the faded frock of an oldtime traveling preacher. The resemblance soon ends. A thin neck is streaked with white, terminating in a long, heavy beak, all of this capped (and here the bird guide fails) in an impossibly brilliant, iridescent scarlet crown. He clutches his post and grooms vigorously. I fear he may have more of a meal crawling beneath his feathers than in that barren stump to which he clings. The shining red crest flashes in the sun as he pokes the long beak down his back.

Moments later a redwing lands without warning on the top of the stump, and this so startles the woodpecker that he falls off backward into the air in a great thrash of feathers. But the big wings catch him and take him easily across the pond in the loopy style of woodpeckers, up to another stump. Here he peers this way and that as if to determine what it was that gave him such a fright. Then as if to reassure himself, he suddenly attacks the post with a violent *rat-a-tat-tat*. It reverberates loudly around the pond like a burst of machine-gun fire. I'm glad I had warning, or

I might have fallen off *my* perch. How exciting it would be if he nested nearby, but I do not see him again.

A. C. Bent says that the original name of this bird was "logcock," but that an Englishman, Mr. Latham, far removed from the local scene, bestowed the name "pileated," from a Latin source. In the opinion of Mr. Bent, who greatly preferred the original, "upon this splendid creature a dull piece of pedantry remains hopelessly fixed."

Bent tells us that the female drills a hole as much as twenty inches deep and six inches in diameter. She digs this huge cavity in dead soft wood or hard, even occasionally in living trees, over a period of two months in the spring, and does it afresh every year, apparently to assure a clean nest free of parasites. What keeps her brain from turning to jellied soup under such prodigious pounding?

Another interesting question concerns the pileated's, like the flicker's, reported fondness for ants, even "big black timber ants." The question is, how do these birds keep from getting their tongues bitten? Would you stick your tongue into a freshly excavated colony of big black timber ants? Quite unthinkable. Questions like these need answering.

A great crested flycatcher has been hunting up and down the shore. I identify him by his lengthy rusty tail and yellowish undersides. The crest is almost nonexistent, however, being only a certain rough lay of the feathers about the head. I would have named him a great rough-headed flycatcher. He calls his loud clear *whirrp* which descends in pitch at the final note. He has an odd way of cocking his head, somewhat like a flicker, as he looks about him for his breakfast. And like the flicker he nests in holes, our only flycatcher to do so.

His breakfast is not hard to find. On a ramble to the north end of the pond this morning I got an inkling for a few minutes of just how rich is the daily banquet set before him. Soon after sunrise, the sun's rays, streaming in from east to west, caught the flying insects of the pond in much the same way that a sunbeam pouring through a window occasionally lights up the dust motes in the air of a room, leaving one hardly able to breathe for fear of choking, so thick is the dust revealed. And so it was over the pond. The insects danced in such a swarm in the clear morning air that I wondered how the swallows protect their eyes as they streak through this teeming host. Somewhere I read that in a square mile of sky can be found more insects than there are people on earth. I don't believe this, of course (who could count them?), but the summer air provides an endless, bounteous smorgasbord for those agile enough to snatch their share.

A First Visit

JUNE 15. A quick look around the pond to see what's about this morning. At first, nothing. Then directly in front of me there is a flash of color that drops off a branch and disappears into the grass below. Nothing for several moments, then a flurry of wings and the visitor flutters to the top of Near Stump. There is no mistaking it. It is a bluebird. He perches facing away from me, showing only the brilliant blue of his back. In a moment he turns, and there is the bright orange chest and white undersides. Now he flies over to a branch close above me. We inspect each other closely. This is my first close look at a bluebird. Am I his first human?

Yellow Petals

JUNE 16. Cheerful, joyous, merry, lighthearted, spirited—observers wax unabashedly lyrical when describing the goldfinch. In his brilliant spring garb the body of the male is as dazzling yellow as any canary. The backs of his cap and wings and tail, though touched with white, are as glossy black as the coat of the redwing.

The habits of the goldfinch are as appealing as his attire. They do not nest until late in the season, from mid July to late September, because, it is said, they insist on waiting for thistledown to appear with which to line their nests. Once the tightly woven nest is made, and rimmed with spider silk, the eggs are deposited and incubated solely by the female. She in turn is fed with faithful attention by her mate, which regurgitates little clumps of seeds for her from his crop.

However, it is only June; they're not nesting yet, but they're beginning to pair. Earlier in the season flocks of goldfinches can often be seen tumbling their way through bare orchards in joyous, erratic flight, like small yellow petals driven wildly before the wind. Today, in a quieter mood, a pair flashed up into the branches above me. They have just dropped into a sunlit spray of pink mountain laurel, flitting lightly from branch to branch.

Tight Fit

JUNE 17. A tree swallow is again poking into the hole in Near Stump. He ducks his head in repeatedly, but does not enter. He finally flies off, returning several times during the morning to repeat the performance. I once set up a small birdhouse with an entrance hole slightly less in diameter than the one and a half inches normally required by swallows, and an anxious pair went through the same head-ducking procedure for two days before they entered. The birdhouse, of course, had room enough in which to turn, so I assumed their problem was being certain they could get through the hole without being wedged tight. Near Stump has a worse problem. The hole is one and a half inches in diameter, but the inside chamber is less than three inches in diameter, and only about four inches high (I measured it last winter standing on the ice). Once through the hole the swallow must be able to turn around to come out. A bird is sleek in only one direction. His feathers make him a veritable bundle of barbs in the wrong direction, and he could not back out of the hole once partially entered. Therefore, perhaps, the extreme caution about entering a hole that might be uncomfortably snug. The hole in Near Stump is tested repeatedly during the season by the swallows, and often defended against other seekers of holes. But it is apparently too small, and no swallow ever nests there.

End of *Che-beek*

JUNE 19. The bluebird is back. He just flew over to Far Stump, where he took a quick peek into the hole and hastily backed off, presumably having met a hungry bristle of little swallow beaks. Now father swallow comes streaking back and the bluebird scurries off. A few minutes later, nothing daunted, the bluebird returns for another peek, and this time is vigorously chased across the pond by the swallow.

A frantic fluttering nearby announces a new arrival to the pond. The least flycatcher has found his mate and is apparently consummating his nuptials in midflight. The birds are locked together, tumbling in the air in a most precarious manner. They separate suddenly with loud squeaks, and dart for the forest. The familiar *che-beek* with which the little male has been calling incessantly for weeks is finished. So persistent has been the call that I have wondered how he has had time to feed himself. An hour later he is all business, flitting silently from branch to branch, tugging at bits of bark so vigorously that several times he falls off a limb. Perhaps I shall be able to locate the nest site and observe these newlyweds at work.

Beauty

JUNE 21. It is the day of the summer solstice. As if to celebrate the event, the sun sends its first rays streaming into the pond, lifting the mist in wraithlike swirls, leaving clumps of water lily leaves gleaming wet and dark in the white luminescence. One can imagine peoples of an earlier and more reverent age kneeling in awe to greet this daily miracle of light and beauty, dreading always a time when the sun might not arrive, and the earth plunge into endless darkness.

Light is the source of all, and its first offspring is beauty. The beauty of this pond is its most moving—and unmanageable—aspect. Like a gossamer cloak flung over the valley with these first golden beams, shimmering, bodiless, unsubstantial, beauty eludes the grasp, frustrates description, and seems of no use whatever. It cannot be draped on the back to guard against the morning chill, or boiled in a pot, or driven or ridden or eaten, or fired at an enemy, or used in any sensible way. As a survival mechanism our susceptibility to beauty would seem to be a bust, a burden that lessens our efficiency and increases the inflation rate.

Are animals touched by beauty? They do not speak, and cannot tell us. As a young boy—an inarticulate young animal myself—I spent much time in a broad rural river valley living on a dirt road that connected several farms that thrived on the rich bottom soil. My most poignant memories are of walking that dusty road under a summer sun, kicking the hot sand, watching sparrows peck in the bordering gardens, listening to the crows calling from the far end of the valley. I was not self-conscious then, as I am now, and did not comment to myself on the beauty of the day, but only experienced it, and remembered it, so that I never hear a crow call in the dawn without a sharp memory of that scene—and a powerful longing. Surely, say, a young polar bear, captured and taken from the wild beauty of the tundra, and placed in the zoo, thinks not consciously of the desolation of his new surroundings, but like the child grown may be deeply moved by the memory of the beauty of his homeland.

Is love of beauty a cultural development of modern man, an outgrowth perhaps of his growing appreciation of the power of order and symmetry? It is unlikely. It is now believed that Neanderthal man, he of the brutish brow and powerful muscles, laid out his dead in a bed of flowers. A Navajo ends each stanza of a chant with the statement, "I walk in beauty," repeating a refrain carried down through generations from his Stone Age ancestors. The American Indian probably had a sensitivity to natural beauty that considerably surpassed that of his "civilized" conquerors.

Yet even if the appreciation of beauty is instinctual in man, its importance is not universally accepted. A naturalist acquaintance who is deeply involved in a study of the complex world of certain insects once said to me, "I'm not into the beauty business. I like to deal in facts. I like detail. My wife," he added, almost in afterthought, "is into beauty."

Staring at this pond is not unlike gazing, enraptured, at a superb painting, where the scene itself invokes deep emotion and "meaning." There is no need to understand the canvas maker's art, or the composition of pigments, or how the frame is joined, or indeed to know the geology or biology or culture of the scene portrayed. To know these things may enrich our experience, but the picture can be known and treasured for itself. It has its own reality. And so with this pond. He who probes deeply into the myriad life forms thriving here, and yet fails to lift his head and see the surpassing beauty of the whole, has denied himself an exaltation of the mind and spirit.

Yet beauty is not without a potentially and profoundly practical use. We *are* in the "business" of beauty. From local garden club to conservation commission to national wilderness groups we struggle to save the beauty and purity of our environment. Perhaps our inexplicable love of beauty, nurtured and fostered among the populace, will stand against our terrifying penchant for contaminating the earth, and become an increasingly practical and powerful protector.

The Odd One

JUNE 23. While watching a bluebird on a stump down in front of me, I kept hearing a low guttural chucking. The bluebird sat on his perch scratching himself and paid no attention to this strange noise. Soon appeared a ruffed grouse, which we erroneously call a partridge in our part of the country. He was mottled brown, turning to soft speckled gray underneath. Two distinctive mark-

ings were a broad black band across the tail and a clump of brown spiky feathers that stood straight up on his head. He was extremely alert, as is perhaps any bird that spends much of his life on the ground. His head jerked quickly up and down and then sideways with his beady brown eye seemingly fixed on me. The chucking became more rapid, sounding even angry at times, then was suddenly replaced with a strange, low, mewing sound. At intervals, the ruff flared out around his neck in an apparent gesture of defiance, but he neither drummed nor spread his tail. He continued this odd bobbing and weaving for almost half an hour before finally disappearing behind his log.

Suddenly, without warning, the great bullfrogs unleash their chorus. It is 9:30 A.M., but the morning mist still hangs eerily about the pond. The deep bleating of the frogs is so loud it defies description. They sound like alligators in a Georgia swamp.

He Kept His Head

JUNE 24. The bluebirds have moved into Far Stump. The swallows have gone. I would like to have seen the young ones fly for the first time. Did they all fly from the nest on the same day? A week ago, a few miles away, I approached a phoebe nest too closely. The young birds suddenly flushed out of the nest all at once and flew quite competently up onto a nearby shed roof. But this could hardly happen to tree swallows. They have to come out of the hole one at a time. Do they fly to different parts of the pond? Do the parents keep track of them all? Or are they only able to locate and feed one or two? One thing is obvious, baby tree swallows are locked in their holes for the duration of their preflight days, up to a month. They cannot flutter their wings in practice flight. The day they stretch their wings to fly, fly they must! A further hazard, according to one expert, is that the wings of baby swallows are so long, and their legs so short and weak, that they cannot hop into the air and achieve flight from the ground. If they fail to land on a branch on their first and crucial flight, they are probably doomed.

The two bluebirds are both down below me now, each perched on a dry stick over the marsh grass. I watch the male. He seems a very gentle bird. He perches quietly on his stump, peering about with surprisingly slow head motions. Now he flutters off his stick, hovering briefly over the grass before dropping down for a small fly, then back to his perch. This dropping down into the grass is typical of bluebirds.

My wife tells a revealing story of these birds. We have a pair nesting in a hole in one of the apple trees at the farm. One day she found the male trapped in our apple shed. Most birds when caught in a shed or barn fly frantically at the windows, as everyone knows, until sometimes they are so stunned and exhausted that they can be picked up to be released outside. This was not the case with our bluebird. Presumably having tried the windows and found them unpenetrable, he simply hopped about the shed in short flights, landing on any convenient perch. As my wife watched, he flew down to the back of an old chair in the center of the shed and perched there, calmly looking at her. When she walked over and opened a side door, he sat another moment as if to consider this new possibility, then hopped off his chair and flew out. He flew on a brief circuit of the nearby trees of the orchard, then returned to the tree nearest the shed door, and (she swears) perched there for a moment looking at her with his bright black eye. Only after this did he fly off. It is this calmness of demeanor, this apparent intelligence, this seeming rapport with his human observers, as much as his beautiful color, I think, that has so endeared the bluebird to man over the centuries.

Death Watch

JUNE 25. I have for the past half-hour watched a fox hunting along the far side of the pond. He first appeared trotting lightly along the top of the wall, stopping periodically to peer attentively into the grass. He is a very poor-looking specimen, with protruding bony ribs and a tail as thin as a piece of rope. He is a far cry from a healthy fox with its rich red coat and magnificent brush. He almost surely has the mange, which may prove fatal to him before the season's out. But he still has the light, alert way of foxes. Now he is down in the deep green grass with just the black tips of his ears showing. Suddenly he takes several quick, floating jumps into the air, head held high, scouting the grass immediately ahead.

This is no place, however, out in broad sunlight, for a sick fox to be hunting. I walk down to a small tree at my side of the water to get a better look. His eyes are strangely dull and he seems not to see me; neither does he note a dark shadow that suddenly glides ominously across the marsh behind him. It is that of a large brown hawk circling the pond in narrowing and descending circles to study the antics of this strangely unafraid fox. But he that can see a mouse at a mile does not fail to see me too. The hawk and I stare at each other, and at death, under the baking sun. The fox, dimly aware now of the mortal enemies gathered at his back, stops and peers about him. The three of us are caught in a web of harsh tension, and except for the slow circling of the hawk, no move is made. Minutes pass. The fox is spared this day, however. The hawk finally decides on discretion, flapping his great wings and lifting himself in reversely widening rings until he dwindles to a far speck in the impassive sky. The fox continues his hunt, and I return to my rock.

Birthrate

JUNE 26. The question of the baby swallows is partially answered today. I saw a swallow dive at what I first thought was a young kingbird sitting on a low branch of deadwood at the north end of the pond. But it is not a kingbird; it is a young swallow crouched with his back to me, greyish brown in color. He suddenly flutters his wings in a funny little excited motion and stretches out his head with mouth opened wide. Exactly two seconds later an adult swallow streaks at him out of the sky like a deadly arrow, slams to an abrupt, hovering halt in the air above him, plunges something deep into that hungry mouth, then drops off in a marvelously graceful swoop, off after another morsel. This is repeated several times a minute for several minutes. Then two adults perch in the dead branches above the baby and preen themselves. No, there are three adults! Why is there only one baby? Have the others come to sad ends? Why are there three adults? Is the third a helpful "aunt," perhaps one who has failed to bring her own brood to puberty?

Suddenly the young swallow flies. One of the adults shoots off in swift pursuit, catching up and following the baby's erratic pattern almost move for move only inches away. Then they fly off out of sight. I must see this again. What was happening here? I had only two or three seconds. Not enough. Was the adult helping the baby in some way, or encouraging it, or only following in parental anxiety? As I mentioned the other day, the vulnerability of baby swallows on their first flights surely justifies a strong measure of parental concern.

One can come up with interesting arithmetic concerning the survival rate of the young of a species. I don't know if anyone knows the average breeding life of adult swallows, but for the sake of discussion, let us assume a pair breeds for three years, raising two broods of four young ones each season. That's eight young per season or at least twenty-four young over the three-year period. Assuming that these are normal years and that the swallow population is stable, the adults will only replace themselves during their lifetimes. This means that not more than *two* of the young birds will survive, and *twenty-two* will not! At least 90 percent of the young birds will die. For a remarkable discussion of this sort of thing, read Annie Dillard's chapter on fecundity in *Pilgrim at Tinker Creek*.

However, grieve not, fellow bird lovers; curb not your cats. If all these young birds lived there would be a population explosion in the swallow world that would soon strip the environment of its capacity to support swallows. Consider our human dilemma, in which we have sharply lowered the death rate of the young, tempered the illnesses of the aged, and all this without appreciably lowering the birthrate. The result is a world population now doubling itself at the astounding frequency of every thirty years, and human disaster in the making, in large parts of the world, that seems more and more tragically inevitable.

Scarlet Woodsman

JUNE 30. A pair of scarlet tanagers just came shooting out of the woods in violent chase. Are they mating for a second brood? The male lands in a tree nearby, perched in an odd position, very erect with black wings hung strangely low and loosely at his sides like the muscular arms of a boxer. I think there is no bird in the forest as bright as this. The brilliant scarlet body is almost iridescent, even on this grey day. I rarely see one at the edge of the pond like this. They prefer the deep green of the woods, and if you see one perched in the dappled sunlight of a young maple grove your day as a birdwatcher is made!

JULY

The Diver

JULY 4. I WALKED DOWN the path toward the pond and slowly up over the knoll above my rock. It was a sparkling sunny day. A crisp breeze rippled the surface of the pond, sending off little splinters of sunlight. The tossing grass encroaching on the old pond was emerald. And directly ahead, on a high stump, surveying the glittering scene like a ragged little patch of blue sky, sat a kingfisher. The wind ruffled his feathers and threatened to knock him from his perch as he fluttered to keep his balance. A strange fellow, with a big dark blue head and spiky crest. A broad white band lay under his chin, encircling his neck, meeting almost, but not quite, at the back of his head. The small blue body and stubby tail gave no hint of the surprisingly large wings that he will spread to fly. He kept cocking his head to peer with one eye at the water as I ever so carefully arranged myself on a stump. Suddenly he flopped off his post and dove straight for the water, plunging in, head first, with a large splash. I couldn't tell whether he caught anything, and he winged off across the pond in his loopy style.

I circled the pond with the glasses. No snakes in sight today. Have they left? They were so active a month ago. Have they moved to another part of the pond? The grass is green at the water's edge and much higher now. This is not as good a month as May for seeing snakes.

Ventriloquist

JULY 5. A strange new call greets the visitor to the pond this morning. *Chip twang, chip twang*. It sounds like two birds. First the *chip*, immediately followed by the *twang*. The latter note originates about seventy feet to the right of the *chip*, but follows it so closely and with such regularity that it seems as if only one bird could have uttered such a combination. A few more steps down the path and the caller is suddenly and brilliantly visible. It is the scarlet tanager. He sits on a branch above the path, his wings not hung low as they were the other day, but up normally against the body. *Chip twang*. The head jerks slightly with the chip. Then there is no mistaking it; the throat vibrates with the immediatly following *twang*. I observe it repeatedly. But the *twang*, a note not unlike a short medium note on a Jew's harp, emanates eerily from a point seventy feet to the right of the bird. It is ventriloquial. A moment later the yellow-fronted female flits through the tree and the male streaks after her in a quick brief pursuit. But soon he is back. *Chip twang, chip twang*, and the *twang* is heard, as before, unaccountably seventy feet to the right.

Pointed Arrow

JULY 6. Across the pond is a tall dead tree. Most of its branches have succumbed to winter storms, and several holes bored into its smooth grey hide attest to the years of affection and attention lavished on it by generations of birds. One lone branch juts out from near the top of the main trunk, raised high like an old

petrified arm set in a perpetual wave. Just moments ago a great blue heron flapped his way massively up the pond and glided up to a perch just below the top of this limb. He curls and uncurls his great toes around the branch, and in the deliberate and dignified way of herons settles himself securely. He is in profile now, the long neck snaked into an **S**, the massive beak almost white, with only a touch of orange at its tip. The soft white tracery of his neck feathers falls over his shoulders like an exquisitely embroidered shawl. He stands almost vertically. The long knobby legs, I notice for the first time, are soft grey in color.

Now his head lifts quickly and alertly, the long neck almost straight, the beak pointed skyward in a typical gesture of alarm. I am sure he has seen me. I make no motion. This is only a half blind and my head and shoulders are plainly visible. My binoculars get heavy and my arms ache. Eventually the beak slowly drops, and the neck settles again into the comfortable **S** shape. I had been previously watching two pairs of kingfishers who have been unusually active, fishing vigorously up and down the pond. I crave to watch them again, but dare not leave the heron. Now he grooms himself, poking that great beak down through his breast feathers, now stretching an enormous wing out to one side, the better to scratch out hungry mites. Now he simply sits and watches. Long minutes pass. The kingfishers splash noisily into the water below him. Will he never get hungry? I am tired. I feel like the boy at the zoo who wants desperately to poke the snakes to make them *do* something. Move, damn it, bird. I wave my hand slowly, but he takes no notice. He is asleep, head settled deeply into the hunched shoulders, still standing absolutely motionless in the breeze, intricately and inexplicably locked to the branch through those long legs and gripping toes. I relax and rest awhile. Now he wakes, and in a beautiful gesture stretches his head far forward until he looks like a huge pointed arrow. Then he flies off his branch, silently, down the pond and off over the dam. The stately flight of the great blue heron is one of the wondrous sights of the north country.

Mr. Peterson

JULY 7. The pond is quiet. A towhee, back in the woods, sings his "Drink your teeeeea." A song sparrow flits along the edge of the pond, unusually silent. A pair of brownish-olive flycatchers are out on the pond. I despair of identification. Are they olive-sided flycatchers, or peewees, or arcadians or willows? Leasts are easily identifiable by their call, *chemeek, chemeek,* and phoebes by their call and distinctive tail wagging. But who identified all the rest? Perhaps it was done with tongue in cheek. Perhaps Mr. Peterson sat down one day and painted a dozen different variations of the same bird, saying in effect, "Here you are, folks, here are the flycatchers. See one and you've seen them all; you can add a half-dozen different birds to your Life List." If I am ever privileged to meet Mr. Peterson, I shall ask him for a direct and detailed answer to this problem. I suspect he will tell me that if one would learn flycatchers, learn their calls.

Bluebirds

JULY 8. The bluebirds are still nesting in Far Stump. The male is easily spotted around the swamp as he flashes his brilliant blue suit. The bluebirds are the only small birds that seem to take an active interest in me. One of them often lands on a branch close by and looks me over for minutes at a time. Today the male perches on a thin dead stalk about twenty feet away. How exquisite he is. His red bib and sky-blue coat stand out vividly against the green sunlit marsh grass behind him. As he cocks his head this way and that, the sun glints in his small black eyes. He suddenly plunges from his perch in a quick, spirited pursuit of a yellow butterfly. He misses, gives up, and the butterfly flutters unconcernedly away.

It has been estimated by ornithologists that the population of our eastern bluebird may have shrunk a dramatic ninety percent in the past forty years. General loss of their favorite habitat, open fields and orchards, partly accounts for the decline. But rivalry for nesting holes by the ubiquitous house sparrows and starlings, both originally imported from Europe, has been a major factor. Here at this pond we see what may be an adaptation to the problem of competition with more aggressive birds, tree swallows in this case. The bluebirds simply wait their turn to use the hole. Our bluebirds did not start nesting in Far Stump until the latter part of June, after the swallows had departed.

Territory

JULY 12. One phenomenon often puzzles me. Today I walked down a small side path on the way to the pond when suddenly I was brought up short by a brilliant little bird in a patch of laurel. He had a bright yellow cap, black eye streak, and white chest, and was easily identified as a chestnut-sided warbler by a jagged streak of chestnut down each side of his breast.

Very excited at my presence, he chirped repeatedly, flitting from branch to branch, fluttering his wings as if it entice me to a chase. Ah, I said to myself, it should be simple to find this nest that is being so vigorously defended here. But ten minutes of searching produced no nest. I had my camera and attempted a picture, but he circled so closely that he was constantly within the fifteen-foot minimum focusing distance of my long lens. Eventually he flew off.

Two hours later I returned—and found no warbler. A visit the next day produced the same result. And so with successive visits. This is the question: Why did he defend this area so vigorously where apparently he was only a temporary visitor? I have seen other birds spiritedly stake out territories, only to find them gone the following day. Did he have a young one hidden here? Or was he only a lone and mateless male, subject to sudden but aimless impulses of the territorial imperative?

Sixty-Mile Call

JULY 14. I have repeatedly heard in past days a very loud clear call, a description of which I can now quote from my field guide: "A rapid succession of very high clear notes and trills" lasting about five seconds. Today I located its owner. He is a small, chunky, brown bird with an absurdly short tail kept bolt upright whenever he is perching. He is as bold as a king's jester, and has just skipped along a dead branch not ten feet away, observing me brightly with one black eye. He is a winter wren, the smallest of the wrens, but surely their most complex and persistent singer. What extraordinary energy in his call. It would be enormously interesting to correlate the sound energy produced by this small body to that produced by other sources. I estimate that I can hear this wren clearly from a distance of a hundred yards. How much does he weigh? Ounces only. No figure is given in the book, so let us guess three ounces. He can project a clear call three hundred feet. That is a hundred feet per ounce of weight. By contrast, I weigh three thousand ounces. To match this wren I would have to let out a bellow that would carry three hundred thousand feet—or sixty miles! (My children say that on occasion I have done this, but this of course is only a much exaggerated memory of a perhaps somewhat overexcited, but well-deserved, admonition). Is there any sound in the world short of an atomic explosion or an erupting volcano that could carry so far? And be repeated four to six times a minute, hour after hour? This wren is simply unmatched in power. His energy output is incredible.

A Chancy World

JULY 16. One extraordinary thing about this pond is the low level of fear, or apprehension, shown by its inhabitants. Most of them are being constantly and mercilessly hunted by one or several of their neighbors. But in their daily vigorous rounds, though often wary, they rarely seem to show a neurotic concern for this. The frog I am watching, which sits so calmly and openly on the mudflat waiting to snatch a fly with a lightning thrust of his tongue, is in turn stalked by the great watersnake, the hungry bass, the heron, and the red-shouldered hawk. It is not that he is immune to fear. I have watched a large bullfrog, passed closely by a blacksnake, clamp his front feet over his eyes and press his face into the mud in an attitude of such dread, such quivering terror, as to blanch the observer. No, it is not that he cannot feel fear. I think rather that between frights he cannot remember the past, and thus imagine a future, in which he might be eaten. If he could, he would spend his days skulking under the shield of a lily pad, in the shallowest of water, too fearful to live a full frog's life. Nature has not provided him with imagination for his protection. Rather she produces his kind in such abundance, from egg to pollywog to frog, that all the hungry predators of the pond cannot consume him entirely; a few of him will survive to breed and start the cycle again.

It is that creature with the most vivid memory, and the resulting imagination, man, who is the most heavily weighed with the burden of fear. Memory, the source and repository of culture and beauty and creativity, is also the arsenal of fear. It is a storehouse, the dark side of which should be opened on occasion only, to be searched for the lessons hidden there, then firmly closed. Memory of horrors past, constantly reviewed, weakens the will and stultifies the spirit. Fear fills our mental hospitals, pits nation against nation, induces savagery, and seems destined to bring us ever closer to the Holocaust. It's a chancy world, but man's overuse of fear as a tool for survival appears increasingly to diminish rather than enhance his prospects for racial longevity.

Worry, distrust, suspicion, anger—it seems not nature's way to fill her creatures with these fearful progeny. Their muted presence here is a major contributor to the appealing healthiness and vibrancy of this pond.

An Unexpected Service

JULY 19. It is the beginning of a grey day. The pond is quiet—the swallows have left! They've raised their broods and have vanished. What a difference this makes to the general vibrancy of the pond. However, all is not lost. A kingfisher loops in with his chittering call. One of the kingbirds suddenly dashes in violent pursuit of a small grey flycatcher and drives him off. The little male bluebird appears on Near Stump, puffing out his whitish belly and red breast and vigorously scratching himself. A thunderstorm rolls ominously in the west. The bullfrogs like this weather, and their croaking echoes hollowly up and down the far shore.

Now the storm rumbles closer, and the grey cloud cover blackens. Suddenly the storm booms sharply overhead, and the first large drops spatter the pond like a ragged volley of gunfire. I huddle under my umbrella, a good protective device for thundershowers, I hope. The birds have flown for cover and no one stirs. The darkness deepens, and I cower. How frightening must be the thunderstorm to primitive peoples, the gods seeming perhaps to be lashing the earth in fury at some imagined insubordination! Now begins the main assault. One crashing explosion follows another with breathtaking violence. The water of the pond is churned to a frenzy by the pelting rain. The trees bend low before the thrashing wind, and the downpour deafens. The thunderstorm is one of nature's most awesome spectacles; all life quails fearfully beneath its fury, and I shiver nervously at the brightest flashes of light, anxious that the following crack of thunder be not too close. The storm is fast moving, however, and soon lets up. It grumbles its way off to the east, sounding, in its retreat, like a mighty elephant with a rumbling stomach.

The storm is followed, in sudden relief, by a gentle shower. Overhead an excited peeping greets the soft rain. In a large white pine a troop of chickadees is doing something strange, and I watch them. They are plunging into bunches of wet needles hanging at the far ends of the branches. Then, sopping wet, they flit back to the bare boughs near the trunk of the tree and shake themselves so vigorously that their feathers become a blur. Then a quick grooming and back out for another plunge into the dripping needles. I cannot be mistaken. They are literally taking baths in the wet foliage and enjoying themselves as hugely as only chickadees can. The mighty storm has provided a small unexpected service.

Chickadee Friends

JULY 22. There is a certain repeated delight in being a neophyte bird-watcher that must wane to some degree as the years pass and one becomes more experienced; this is the fact that nearly every trip produces a new bird. Today it was a pair of small brightly marked birds that seemed to be traveling with a flock of chickadees. They stayed lower in the trees than the noisy chickadees, and acted like creepers, running up and down dead branches, sometimes flipping completely around a lateral branch. They were prettily streaked in black and white from crown to tail, and at first I was sure I had discovered a new species of nuthatch. But of course I had not. A quick check in the guide identified them as black-and-white warblers. They in their quick lively manner, and I in a heavy-footed thrashing through the laurel, followed the chickadees for a spell before we lost sight of each other.

An hour later: the chickadees have just arrived at the pond, a good half-mile from where I first saw them down the trail, and to my surprise, the warblers are still with them. Do the chickadees in some way provide comfort or protection to the warblers? I have found that other birds often travel with chickadees, sometimes several species at a time, and I have learned to look closely at a flock of chickadees to see who their guests might be today.

Two yellow-shafted flickers sit high in a dead tree across the pond, *whicka, whicka, whicka*. They are in a quiet mood for flickers, who more often utter their long staccato calls. The male is preening himself, stretching one wing wide to the side like a pale yellow, translucent fan. He carefully nibbles along the base of the feathers, where no doubt dwell small itchy mites. He bobs and swings his body in the odd way of flickers, showing his black throat spot and the bright yellow underside of his tail. Flickers are woodpeckers, but more than half their diet is made up of ants foraged from the ground and picked up with long sticky tongues. They are always identifiable by their white rump patches that flash as they fly up the trail ahead of the hiker.

The Tonic of Wildness

JULY 25. A great blue heron flies over this morning. He does not land, and I am reminded of a problem. As I have mentioned, I have had, until recently, a small half blind of faded burlap stretched over sticks propped along the front of this rock on which I sit, concealing me to the shoulders. This provided enough cover so that the big herons often landed less than one hundred feet away and stalked along the marshy edge of the pond without alarm. But my blind was removed recently by members of the conservation commission that protects this area. And they were right to do so. The blind was a shabby little man-made intrusion on this exquisite pond. And so too, if I may mildly retaliate, are the commission's own heavily planked bridges over the streams and their neatly color-coded trails. One yellow blaze stares at me now from across the pond like an off note in a Brahms sonata.

Even the most well-intentioned of men have difficulty keeping their hands off wilderness. It is our nature apparently to grasp and control and modify. In the process we destroy the very wildness we crave. Thoreau said, "We need the tonic of wildness. . . . We need to witness our own limits transgressed, and some life pasturing freely where we never wander." There is an obvious ambiguity in this most famous and popular pronouncement. We cannot witness life pasturing freely where we never wander. To witness is to be present. And we will wander! There is no stopping us. So the problem becomes one of compromise: how to be there, but be as inconspicuous as possible.

The first and major problem is one of numbers. There is no sight that illustrates this dilemma so poignantly as those pictures we see of a pride of lions in Kenya's beautiful Nairobi Park surrounded by a fleet of zebra-painted minibuses. No wildlife pastures freely here. The scene is man dominated, and wilderness has fled. As in our own national parks, where the human population outstrips the wildlife it has come to see, where roads are punched through to the most solitary recesses of the back country, the "tonic of wildness" eludes us.

The ideal witness is a witness of one. Unlike the music of a symphony orchestra, which needs our audience, the music of the wilderness, which at its profoundest pours like a mighty cataract through the heart of man, is diminished and muted in direct proportion to the numbers and obtrusiveness of its human observers. If you would drink the deepest draught of this music, travel alone. Yet even then you will not hear the final note, for the wilderness never sings so powerfully as in the absence of man.

The paradox is unavoidable: man the observer becomes man the despoiler by his very presence. What can be done? We can only aim at the ideal. Severely limit the numbers of us who can visit our most treasured wild places, our national places, at any given time. Insist that we reserve a place in line and wait our turn if need be. Divest us of our vehicles and radios and bright clothes. Require that we take instruction in how to observe without dominating, in how to move, how to sit, how to see. Teach us, so far as possible, not to stand apart, but to become a part of the wilderness. Demand of us some sacrifice, some hardship, some preparation as the price of witnessing our wildest places. Is there any other way? By our numbers, by the very vibrancy of our presence, we destroy the peace we seek.

AUGUST

Ballet Dancer

AUGUST 7. A SOFT WHOOSH of feathers to my left alerts me to look that way just in time to see a great blue heron land at the edge of a watery patch. I catch a glimpse of the broad dark bands along the rear edges of his wings, which disappear completely as the wings are folded into the slate-blue body. The heron stretches his long neck to peer ahead, then starts a slow stalk across the marsh. Each leg is pulled from the water, then the foot, with wide splayed toes, and foreleg are placed flat on the surface and pressed down slowly until footing is found. The neck folds to a sharp **S** shape as he stares intently ahead. There is a knobby spot at the **S** curve of the neck, typical of these herons, as if he had dislocated a vertebra.

Now something catches his attention and he hunches his body, sinking nearly into the water. Now the strike! The beak plunges deep, head and neck entirely disappearing into the water. A tiny fish squirts into the air ahead of him. He missed. He resumes the quiet stalk, pecking occasionally into the grass, gulping what he finds; probably insects. Now another crouch — a deadly posture with neck snaked and huge beak poised. A deep strike! A flurry of minnows breaks water ahead of him, but this time he comes up with a small fish dangling from one side, about halfway up the beak. He shakes the water from his head, neck feathers fluffing out thickly, chomps briefly at the fish to line it up, gulps and swallows. When he stabs at a frog on a mud flat, I think, the strike is lethally accurate. But when he plunges, as now, deep into a school of fish, I think he simply snaps on his way through the school trusting to a lucky catch. Or he may not even snap, thrusting only with a partially opened beak, any unlucky fish being wedged into the long inverted **V** of the open mandrels. Either method would account for the fish being caught by its tail halfway up the beak. I estimate that fewer than a third of these plunging strikes are productive.

The grace of this bird is hard to describe. He is facing away from me now, legs delicately crossed as he twists and peers intently far to the right, neck arched over the water, the sculptured grey body leaning to the left in counterbalance. No ballet dancer could dip in a more exquisite posture. He is a youngish bird, I think, with only a touch of the white embroidery down his chest. He spreads his great wings to flap a few yards ahead in the light air, and floats down so buoyantly that he seems to have divorced himself almost entirely from the tug of the earth. He disappears finally into a channel in the deep grass and I see him no more.

Prodigious Fisherman

AUGUST 14. A great blue heron flew up from the edge of the pond this morning when a twig cracked under my foot. He was only mildly alarmed and flew slowly down the length of the pond without gaining altitude, then circled back and flapped his way over my head, obviously checking to see what the disturbance was. He flew off quietly. Great blues when sharply alarmed always fly off with powerful strokes, gaining altitude rapidly and uttering deep protesting croaks.

I sat for half an hour at 8:30 A.M. and not a bird moved. Only the droning of the insects indicated life on the pond.

Then *loopedee loopedee* along came a kingfisher. He glided to the top of Far Stump, sat for five minutes, launched into a downward glide at 45 degrees, and plopped into the water with a loud splash. A small boy diving into the pond would not have made more noise. Then he flew up to a tree, and quickly dove again. Now he flew to the top of Near Stump, where I could observe him closely. He uttered a series of his long, chittering calls, flicking his tail up and down and cocking his head comically from side to side. His absurdly spiky hairdo, brushed up stiffly from his forehead, looks as if it has suffered the ravages of an amateur barber. His long, heavy, dark beak completes the feeling of massiveness about the head. For all this, he is beautiful. He is a male, lacking the lower brick-colored belly band of the female. The guide book describes kingfishers as "perching motionless in the open." Kingfishers perch on stumps, branches, sometimes on rocks at the edge of the pond, but they are never motionless. The head is continuously cocked from one side to the other and the tail flicks as the bird watches for the least sign of movement in the water.

Somewhat later I saw this bird gorge himself into a state of near stupefaction. Remarkable behavior accompanied this gourmandizing. The bird flopped off a high branch across the pond and hit the water with his usual noisy plunge. I had never before observed a kingfisher holding anything in his beak after one of these dives, and had always assumed that he swallowed immediately whatever he caught. This time, however, he flew back to his branch, and I saw that he had a large fish in his mouth, much too large, in my estimation, for him to swallow. This fish wriggled vigorously, and the kingfisher shook his head vio-

lently. Then he suddenly flew down and landed on a long silvery-grey log lying next to the water's edge.

Then ensued one of the most astonishing scenes that I have witnessed in the bird world. He held the quivering fish firmly across its middle, raised it high, cocked his head, and swung it smartly sideways, driving the head of the fish into the log with a resounding smack. He raised his head and again vigorously whacked the fish against the log. The small slapping sounds came to me distinctly from across the pond. Each time he raised the fish so high that he seemed to be stretching to his tiptoes; this of course gave him maximum leverage for the powerful blow to the log. After he had repeated this bludgeoning several times, I began to count. At the count of twenty the fish was still wiggling. At the count of forty the fish seemed to be dead. Yet the hammering continued at the rate of a blow every three or four seconds. The blows were nearly all to the right and a wet stain slowly spread on the light grey of the log. Still the battering went on. Eighty blows. One hundred blows; one hundred twenty-five. The fish was as limp as a rag and discernibly pink and raw about the head.

What was the purpose of this bashing? Several of my books claim the kingfisher does this to kill the fish. This seems nonsense. This fish has long been dead. Is it done rather to break down the bony structure of the head, to crush the skull and rib cage of the fish to a gulpable size? Perhaps it is not swallowing that is the problem, but rather that the large bones of the fish must be crushed before they will pass through the digestive system. Perhaps the hammering and pulping of the flesh of so large a fish helps to hasten its digestion.

One hundred fifty blows. No letup yet. So far all the blows had been to the right, but now there were a few to the left, and a stain on the log began to show on that side. The blows were as hard and vigorous as ever. Finally at the count of one hundred sixty-nine—and that was at least ten blows short of the total, due to the late start of my count—the kingfisher raised his head and in two or three gulps swallowed the fish.

He flew back to the top of a stump and sat quietly for an hour. Then, unbelievably, he started to fish again. What a prodigious appetite! How could he conceivably contemplate another such meal within the hour!

A month ago he would have returned, I think, to his nest, set at the end of a long tunnel dug into the edge of a stream bank. There he would have regurgitated a nutritious banquet before a hungry group of near-grown young. Now, however, more likely he is building a heavy layer of fat about his chest, "larding" himself, as birders say, storing energy for winter. He may migrate as far south as Florida, or he may winter along more northern coasts wherever the streams and estuaries stay free of ice.

A Touch of Spring

AUGUST 24. Today on the trail to the pond I notice again the phenomenon of grouping. A small flock of chickadees is busily scouring a large oak tree. Normally I give them a casual glance and pass on; but having learned of the seeming affection held for chickadees by other birds, I sit down to watch. Immediately, what looks like a moulting scarlet tanager, with brilliant red body bespattered with yellow, flits through the tree. Now a small warbler darts in, moving even more quickly and erratically than the chickadees. He is one of the confusing fall warblers, gray brown on top with a soft flush of yellow underneath. No wing bars. A white-breasted nuthatch scampers his surefooted way down the oak, and out onto a branch, and under the branch, his needlelike nails clutching bark so old and loose that I wonder that he does not fall. His hops are so quick that it seems as if he must be moving both feet at once, an act that would seem quite impossible when upside down. I watch as closely as possible, but cannot tell. Now a jay passes through briefly, as if to check on this little localized commotion.

The pond itself is almost devoid of movement on this August day. This is the dry season. The mosquitos and blackflies have gone, and with them have gone those that feed on them: the swallows and flycatchers, bluebirds and kingbirds. Where do they go at this time of year? The pond has receded to a small pool near the outlet, and the only flycatchers left are the frogs that sit silently and expectantly on the surrounding mudflats, awaiting an errant dragonfly. The water lilies have bloomed and spent themselves and have crumpled into piles of brown debris on the mud. Even the turtles seem spent. One is perched on his breastplate on a small rock, all four legs hanging loosely and awkardly from his sides.

But there is still new life. The grasses, the wondrous emerald green grasses that have provided so many hours of visual pleasure, richly fed by the thick nutritious muck of this old beaver pond, continue their inexorable takeover. Spreading over the newly exposed mudflats encircling the diminishing pool is the most extraordinarily delicate flush of green. A touch of spring in the waning days of summer.

SEPTEMBER

The Baiting of a Hunter

SEPTEMBER 12. THE POND IS A PANDEMONIUM of jay calls this morning. Interspersed with their screeches and buglings is an occasional sharp, raspy call with an angry edge. I hasten to my rock to determine the source of the excitement. Down over the old beaver house the trees are alive with excited jays. The object of this attention soon shows himself with a rapid beating of wings and a quick and unsuccessful chase of a jay. It is a small hawk barely larger than the jays. Now another chase. At each sortie of the hawk, the cries of the jays rise to a frantic crescendo. I can only see haphazardly through the trees. Now there are strange "jay" calls that I have never heard. The source of these is soon seen flying hastily across the pond, dodging and ducking, with the hawk hard after him. It is a flicker. He easily eludes the hawk, and the hawk lands in a nearby tree that gives me a good view. This tree is almost immediately inundated by a swarm of jays hopping excitedly through the branches in a constant mosaic of movement. Then one jay hops from branch to branch ever closer to the hawk, until he perches not more than three feet away. The hawk abruptly plunges after the jay in a tight, dizzying pursuit that lasts for several seconds before the hawk gives up. The jay is simply too agile for him to catch. This provocation, and the quick, unsuccessful chase, are repeated again and again obviously creating intense excitement and enjoyment for the jays.

The hawk is of the accipiter family, with rounded wings and a long tail, smaller than the broad-tailed buteo family. He is dark above, light below, and very heavily streaked in brown on the breast. I identify him as a young sharp-shinned hawk, fairly common in our woods. My *Golden Guide*, showing a lamentable tendency to unromantic description, describes his call as a "cackling note." Who could accuse a hawk, a member of that great family of aerial predators, worshiped by man as a consort of gods, of cackling? A. C. Bent describes him as "this bold and dashing little hawk, the terror of all small birds."

Eventually the jays tire of baiting our young hawk and drift off. He sits forlornly in his tree, no doubt hungry, craning his head this way and that. Suddenly he takes off vigorously after a lady kingfisher, identifiable by her rufous cummerbund, that has been fishing noisily and unconcernedly about the pond for the past hour. He chases her nearly the length of the pond, but she easily shakes him off and sails up to a perch on a high stump. He glides into the trees, and I reflect on the fierce necessity, if he is to survive, of his learning soon the skills of the hunter.

Sharp-shinned hawks, like their rarer and larger look-alikes, Cooper's hawks, are bird hunters of the forest. They are designed for quick dodging through trees, and must take their even more nimble prey by surprise. They are said to take birds up to pigeon size. Jays know their danger, and when they find young sharp-shinneds, hound them unmercifully.

Unseeing Hunter

SEPTEMBER 14. Sitting at the north end of the pond, I am alerted by a sharp splash two hundred feet to my right, in the vicinity of the beaver house. To see better, I walk carefully toward the pond until I am standing just this side of an old stone wall that borders the shoreline at this point, and am well rewarded. Standing over the submerged entrance of the beaver house is a red fox. He is very alert. The black-tipped ears point this way and that as he watches the water intently. There are no beavers here now; perhaps he has surprised a muskrat that is living in the deserted house. What a beautiful specimen is this fox, lithe and supple. His rich red coat is tinged at all extremities with black, except for the white tip of his tail. The tail is enormously thick, in dramatic contrast to the slim body and long, slender, black legs. The eyes are thin slits at the top of his pointed nose. He circles the top of the beaver house, peering into the tangle of sticks, and occasionally tries to dig his way in. I suspect he is a young fox that has not yet fully discovered the impregnability of beaver houses.

Now he heads down the shore toward me, tripping down an old log and into the marsh grass. In a few minutes he has reached my position. I can only see the tips of his ears over the wall that lies between us. Ten feet further on he suddenly jumps lightly to the top of the wall and proceeds to trot along the rough boulders. As I have remarked before, the delicate, light-footed way of a fox along the top of a wall is a remarkable and beautiful thing to see. A dog would scrabble and slip; the fox is as surefooted as a mountain goat on a high ledge. He stops, fifty feet to my left, and looks out over the pond. Now he turns and proceeds back down the wall toward me. I am electrified. He does not see me, though I am standing in the open and in plain sight. He seems nervous now. The air is still, but perhaps his nose has detected an alien essence. He trots quickly down the wall, passing twelve feet from me, proceeds a few more feet, then jumps from the wall and lopes rapidly into the woods.

He did not see me! He simply failed to see the unexpected, and he proved an old rule of the wilderness. He who moves is seen; he who is motionless is virtually invisible.

It is interesting to speculate on what wildlife *sees*. I recently read the story of a man who hangs a seat in a tree, and decked in camouflaged suit and blackened face, sits and watches the wild creatures below. I agree with his premise that men and animals tend not to look up, but I doubt that he needs his camouflage and face blackening. I think most wild creatures are not alarmed by shapes, however unfamiliar. A friend describes an incident while hunting in Colorado. Sitting motionless on a log, with a license only for elk, he was approached by a doe deer. She came to within ten feet of him, nose quivering sensitively, before a sudden whiff of him, or perhaps a blink of his eye, sent her bounding away. She simply failed to recognize the shape of a man.

It is said that around African water holes baboons are prized as watchdogs by other animals, and that this is because they can distinguish the motionless shapes of lurking predators, while the

grazing animals cannot. Smell, sound, motion, an unfamiliar aspect in any of these will trigger alarm. But often a simple shape will not. Sit motionless, a little back from the edge of a pond. The wild black duck, obviously from their wanderings familiar with the shape of man, will land in the water before you. Even the great blue heron and the wary fox will treat you as one carved from stone, if in fact they see you at all. And, of course, like humans, wildlife often fails to see what it does not *expect* to see. So I think it was with our fox. We need only sit, immobile, to have the rich pageantry of the wilderness roll before our eyes.

Ducks and a Pair of Visitors

SEPTEMBER 21. Very cold this morning — 45 degrees. The time of the great winging southward of the waterbirds, from far up on

the Canadian tundra, has come. It is the time of ducks, and a sudden loud quacking warns me to be careful. Two days ago I watched a wood duck family slowly feeding its way through the marsh. They had probably summered nearby in a hollow stump near the water's edge. The male had moulted to a uniform brown, but with patches of his earlier magnificent plumage remaining. Still much white under the chin, which will remain, but most striking was his eye, a large, round, sparkling red ruby with a black dot in the center. The female's eye was equally beautiful. It was black, but surrounded by a tear-shaped white ring, rounded in front and tapering to an exquisite point at the rear of her head. She is colored permanently brown except for a wing patch of deep blue showing near her tail. One young one followed, as large as his parents, but with no coloring except for the dash of blue.

Today it is black duck that are giving the warning. Five of them circle alertly directly in front of me. Through the trees I can see an animal swimming out beyond them, and it is this, I think, rather than I, that is causing the alarm. I take a long time to get to my rock, and only crouch behind it. There are many eyes today. Several more blacks are down near the dam and a family of four wood duck are feeding at the north end. A few minutes later a large hawk lands in a high tree beyond the wood ducks. Facing the sun, his front is a shining brick red from his chin to his toes, and out onto his shoulders. He is a red-shouldered hawk, my first. There are several jays about, but they don't bother him. He stays only a few minutes, seemingly unnoticed by the ducks, before flying off.

The pond is tranquil, but there is a rare sight to come. It is announced by a low rippling splash down the shore to the right. A first look provides only a glimpse of a brown body disappearing into the water. The beaver are back, I think with pleasure. Then a small brown head pokes up and looks around alertly, like a

trim periscope. Then another. Both heads go down, followed by a humping of the sleekest brown bodies I've ever seen. They are a pair of otters. Up they come again. Quick and restless. They turn and dive and reappear, exploring a small pool of open water in the brush. They are wonderfully smooth. Not a ripple now. It is as if they were swimming in oil. They have a luxurious, almost voluptuous relationship with the water. They are liquid themselves, so fluid in their motions that one cannot imagine their bones. Three minutes. Four minutes. They are gone. Out through the slipway of the dam, perhaps, although I did not see them go. I hope they'll return, but otters are great travelers, and they may be miles away by tomorrow.

The four black duck (they are not black, of course, only a dark mottled brown with blue wing patches and white underwings) flush into the air, quacking excitedly, as I get up to go. They beat rapidly off over the dam, and out of sight. Then, as I've often seem them do, they return, flying silently above the treetops, peering down to see what it was that frightened them. It is a dangerous curiosity, I think, one that has no doubt put many a fat duck onto a Sunday table.

A Touch of Wine

SEPTEMBER 26. A thin scud of grey clouds drives high over the pond from east to west, the rag end of a heavy northeaster that drenched us yesterday. The water is high, the pond inexplicably quiet. The old grey poles along the far shore are reflected in the pool as if, sixty years ago, when they were only seeds, they had grown both up and down, the image being hardly less clear than the object. Should it not be so with us? The limpid eye, perceiving objects, reflects only images to the brain. But in what contrast to this placid pool are the turbulent surfaces of the mind.

Churned by the daily turmoil of our lives, clogged with decades of accumulated notions, rigid with the fears of an unsure, ever fluctuant world, beset by the fierce demands of ego, it is a miracle that we see at all. Distortion splinters truth and dulls the eye. Calm, calm yourself, my mind. Let the restive waters sleep, that I may see, and hear.

As if to relieve the somberness of the day, an oak has begun to turn color, and extends one delicate wine-colored branch out over the water. The great draining of the chlorophyll has begun, and every leaf will soon show the brilliant colors that have lain dormant and unseen these many months beneath the light absorbing green.

Numbers

SEPTEMBER 27. The most ubiquitous bird of the swamp in June is the swallow. In late September it is the jay. Congregating in flocks (perhaps more aptly called gangs), they call to each other with harsh cries across the water. The swallows, so dependent on the great hordes of insects that rise into the air in the spring, are forced, when their food base collapses, to head south. The omniverous jay, which feeds heavily on larger insects in the summer, happily turns to fruit and seeds and acorns in the fall, and finds enough to survive the winter.

Or perhaps first place for numbers should go to the ducks. Nine black duck passed a moment ago, high over the north end of the pond to my left, winging rapidly along in their serious and purposeful way until they disappear over the far woods. Now they reappear over the south end, banking in sharply to head up the pond. Suddenly, as if on signal, discipline disintegrates. They break formation and come fluttering down, fat bodies swinging back and forth like plump pendulums beneath their wings, widespread feet thrust comically out in front. They hit the water with hard splashes, one after the other. They stay only a few minutes. There is a sharp breeze blowing, and as every duck hunter knows, ducks move around when the wind blows. They soon depart. Of course, I had just swung my glasses too quickly to watch a kingfisher emerge from the water, and might have contributed to the short visit. Black duck tolerate no sudden movements.

Almost a minute later there is a sudden scrabbling behind me, and a low chittering sound. I turn, slowly this time. It is a young raccoon. He humps his way along in a droll, rolling manner, on surprisingly skinny legs, busily digging under rotten logs, until he arrives at a spot less than ten feet from my rock. He has none of the alertness of the fox, and, in fact, seems cheerfully oblivious of the world about him. He chitters, or whickers, almost continuously as he works his way out of sight down the shore. He proves one thing about coons. They are not strictly nocturnal. It is almost 9:00 A.M. and broad daylight, though cloudy.

OCTOBER

Prince of Soarers

OCTOBER 1. A GREAT TURKEY BUZZARD glides low overhead, his wings spread in a broad dihedral shape, the primary feathers at the tips spread so wide they look like fingers. I watched three of them recently circling in a tight spiral for twenty minutes, as if in some kind of buzzard minuet, until they drifted higher and higher and finally out of sight. Ungainly and ugly on the ground, the buzzard in the sky is the elegant prince of soarers, and I never tire of watching him.

Conflagration

OCTOBER 3. Each year one hears a neighbor say, "I think the colors this year are the brightest I've ever seen." The New England turn of colors is so spectacular, so magical, so overwhelming, that one simply cannot remember how it was the previous year. One great, lone maple standing massively in an upland pasture, spared by some early farmer from the scythe, suddenly one day in mid-October is a full blazing scarlet, so totally aflame, so impossibly brilliant, that the mind can fully hold the image only so long as the eye transmits. It is an experience to be repeated each year with fresh astonishment and delight for the rest of one's life.

And so it is with the swamp maples along the north shore of this pond. Silhouetted against a backdrop of dark green pines, they have suddenly swept into a conflagration of orange and yellow and fiery reds that range from scarlet to deep wine. They are reflected in the water channel wending through the still green grasses, and seem illuminated by some flaming inner light of their own. The maples around our swamps turn first. It will be another week before this sight is duplicated in the hills, and travelers will come from hundreds of miles around to see afresh this miracle of transformation.

Weak Feet

OCTOBER 5. A pair of red-shouldered hawks are perched motionless side by side in a dead tree, one slightly lower than the other, at the south end of the pond. They are facing me, and their chests glow in the early slanted sun as if they were carved from red brick. They remain a long while and I fail to see them leave.

I watched one of these hawks in almost the same location the other day. As he waited in his tree, a grey squirrel ventured out onto a log not seventy feet from his perch. I hardly drew breath, waiting for the incipient attack. It never came. The squirrel twitched his way along the full length of the log with impunity and hopped safely to shore. Later I read that the red-shouldered hawk has much weaker feet than those of his heavier and stronger cousin the red-tailed hawk, and because of this limits himself to small rodents and birds and frogs. A grey squirrel is extremely muscular, and can inflict a savage bite when cornered. He is apparently too strong a prey for the red-shouldered hawk, which himself may be attacked by stronger hawks and owls.

Autumn

OCTOBER 10. Autumn is both a difficult and an exciting time for watching birds. It is a time of young birds, indistinct of plumage and sometimes difficult to identify, but exuberant with youthful energy and derring-do—the young blue heron fishing with his parents, for example, which bounces into the air every few minutes on his great new wings; the young jays chasing hawks, frantic with excitement; a pair of skinny-legged young raccoons that hump along out of the woods, chittering lightly, and interrupt their breakfasting long enough to roll in the sunlit leaves in a brief tussle before working their way back into the forest. It is also a time of larger birds. The small hunters of grubs and insects are still here and will stay—the chickadees and nuthatches and a downy woodpecker that is at this moment squeaking loudly overhead as he pecks hollowly at the smooth grey shell of an old limb. But it is the big birds that now dominate the swamp—the hawks drifting through daily on their southward migration; the swift-winged ducks, and the jays and crows and flickers that suddenly in their fall congregating achieve a larger presence.

Six Eyes

OCTOBER 12. Far Stump, which has seen little activity for some time, seems to have spawned its own grey offspring this morning. Three young grey squirrels race up and down its weathered hide, chasing each other around the trunk until one pops up over the top like a jack-in-the-box. They dive repeatedly in and out of the hole that has so cozily housed several generations of birds this summer. Now all three disappear into the hole. How can it possibly hold them all—like college boys crammed into a telephone booth? There is a flurry of fur at the entrance, and a sharp little face pokes out; now just above it, another; now, unbelievably, a third pops out above the other two. Three small faces in a vertical row. Three pairs of bright black eyes peer sharply over the pond. Now they explode from the entrance, one after the other, in a grey whir of flicking tails and needle claws. They show no interest in fixing up the hole for a winter nest, however, and soon scamper off into the forest.

Silent Feeders

OCTOBER 22. Three black duck are the only visible occupants today. They are busily feeding one hundred fifty feet up the pond in the new high water of last week's rain. But the high ground is dry again, and it has been a slow stalk getting to my rock through the thick carpet of crackly leaves and into my seat without flushing them. When the drake looks up sharply, swimming in tight circles with his head held high, I freeze. It takes five minutes for him to settle down. From then on no amount of peering about on my part seems to disturb them. Most of the big hawks have gone, and I despair of seeing much bird life at this late date.

A light patter of rain begins, rattling on the dry leaves, and this is strange because the sky is clear. I ponder this improbability for some minutes, watching my sleeve to catch sight of a droplet. Perhaps it is a light hail coming from unseen moisture in the sky. The patter continues. A plink on my coat. It is a seed. I look up, straight up, until my hat falls off the back of my head. And there in the top of the black birch that towers over this rock is a silent flock of little brown birds, busily picking at the small seed pods hung at the end of every twig. I eat one of the seeds. It is good, though a little woody for my taste. The birds have heavily streaked breasts, long, deeply forked tails, and a touch of yellow here and there. They are pine siskins, similar in action to goldfinch, with which it is said they often flock. They fly off with small squeaks in a wild, exuberantly erratic flight.

Now jays, chickadees, and nuthatches call from about the pond. In spite of strong wings, and warm lands beckoning from the south, they stay to face the fierce rigors of our northern winter. They inhabit a world that is visibly drying out, the earth withdrawing the moisture that will be so subject to freezing in the impending heavy frosts. It is Indian summer, with brilliant days to thrill the eye, but it is the low moan of the evening wind that quickens the pulse with fierce pleasure at the thought of wintry gales to come. I know not how the creatures of the forest will stay warm, or whether they will find enough to eat, but I have no inclination to follow those that have flown to the south.

End of the Season

OCTOBER 24. It is a golden world this morning. On the way to the pond, a grove of tall, straight maples and beech that once shaded this portion of the forest now stands stripped and bare, bedded in its own thick blanket of sun-struck yellow leaves. The light is everywhere. Even the hills, now visible through the trees, glow as deeply lavender as in any sunset. Moments later a sudden gust of wind carries a bevy of small leaves down the length of the pond. They glint brilliantly, rocking and dancing in the sun like a flock of golden butterflies.

This pond has been a place of singular beauty to me. Aesthetically it is a small gem that never fails to delight the eye and ear. It is a place of music and ballet. Even now the breeze rustles the leaves and tempers the warm sun captured here. It is a place of thunderstorms and thick winter ice and the pulsating summer-evening chorus of frogs and birds. It is a small, wild spot that strikes deep chords in the human observer.

Why does such a place produce such emotion? Is it truly a magical place, or is this only a poetical illusion so easily created in the imaginative human mind? It is a mercilessly practical place, populated by hunters and hunted, each life sustained by constant death, a place of intense, enormously energetic pursuit of survival. And it is a place of love, loyalty, courage, beauty, excitement, and adventure. It is all these things, without conscious cruelty or duplicity or premeditated greed. It is an instinctive world, unselfconscious, a tiny piece of the mighty original wilderness from whence we came, and the yearning to return is great.

Sources Consulted

Amos, William H.: *The Life of the Pond.* McGraw Hill, New York, 1967

Bent, Arthur Cleveland: *Life Histories of North American Birds.* Originally printed as Bulletins of the United States National Museum, 1919 to 1968. Reprinted by Dover Publications, New York.

Robbins, Chandler, S.; Bruun, Bertel; Zim, Herbert S.: *A Guide To Field Identification, Birds of North America.* Golden Press, Racine, Wis., 1966.

Terres, John K.: *How Birds Fly.* Hawthorn Books, New York, 1968.

Van Vleck, Sarita: *Growing Wings.* William L. Bauhan, Dublin, N.H., 1977.

Wetmore, Alexander: *Song and Garden Birds of North America.* National Geographic Society, Washington, D.C. 1964.

ERIK BROWN first discovered the isolated pond of this book a few years ago not far from his small farm in southern New Hampshire. He brings to his observations of the natural world the precision of an engineer, the imaginative curiosity of an artist and inventor, and the enthusiasm of a new convert.

He was born in Berlin, New Hampshire in 1923, and during World War II served with the Marine Corps in the Pacific. In his own words, he "mixed bouts of education at Williams College and the University of New Mexico, with a variety of jobs ranging from framemaker and busboy, to disk jockey in Texas, ski instructor in Colorado, and 'roughneck' on oil rigs from Texas to Wyoming."

Thereafter he returned to New England to devote twenty five years to "uncharacteristic close attention to business" and to inventions of machinery and packaging products, for which he received several patents. This earned him an early retirement to his farm in New Hampshire, where he lives with his wife and family, and devotes his time to woodcarving, writing, and his new found vocation of birdwatching.

ELIZABETH ADAMS is a free lance graphic designer with a studio in Hartford, Vermont. While at Cornell University, from which she graduated in 1974, she spent summers as artist-naturalist with the New York State Department of Environmental Conservation. Her illustrations have been published in *Country Journal*, *The Conservationist* and *Dartmouth Alumni Magazine*.

*This book was typeset in Bem, a facsimile
of Aldine Bembo, by A & B Typesetters, Concord,
New Hampshire, and printed and bound at Halliday
Lithograph Corporation, West Hanover, Mass.*